SpringerBriefs in Materials

The SpringerBriefs Series in Materials presents highly relevant, concise monographs on a wide range of topics covering fundamental advances and new applications in the field. Areas of interest include topical information on innovative, structural and functional materials and composites as well as fundamental principles, physical properties, materials theory and design. SpringerBriefs present succinct summaries of cutting-edge research and practical applications across a wide spectrum of fields. Featuring compact volumes of 50 to 125 pages, the series covers a range of content from professional to academic. Typical topics might include:

- A timely report of state-of-the art analytical techniques
- A bridge between new research results, as published in journal articles, and a contextual literature review
- A snapshot of a hot or emerging topic
- An in-depth case study or clinical example
- A presentation of core concepts that students must understand in order to make independent contributions

Briefs are characterized by fast, global electronic dissemination, standard publishing contracts, standardized manuscript preparation and formatting guidelines, and expedited production schedules.

More information about this series at http://www.springer.com/series/10111

Plinio Innocenzi

The Sol-to-Gel Transition

Second Edition

 Springer

Plinio Innocenzi
Department of Chemistry and Pharmacy
CR-INSTM Laboratory of Materials Science and Nanotechnology
University of Sassari
Sassari, Italy

ISSN 2192-1091   ISSN 2192-1105   (electronic)
SpringerBriefs in Materials
ISBN 978-3-030-20029-9   ISBN 978-3-030-20030-5   (eBook)
https://doi.org/10.1007/978-3-030-20030-5

This Springer imprint is published by the registered company Springer Nature Switzerland AG
The registered company address is: Gewerbestrasse 11, 6330 Cham, Switzerland

The International Sol-Gel Society (ISGS)

Dear Readers,

It is a great pleasure for me to write a preface about a book from Professor Plinio Innocenzi. This book is the second edition of "The Sol-to-Gel Transition" in the "SpringerBriefs Series in Materials." This collection of books presents highly relevant, concise monographs on a wide range of topics covering fundamental advances and new applications in the field. "The Sol-to-Gel Transition" is one of the most recommendable books for early career researcher and the new comer of the field of sol-gel science and technology.

The International Sol-Gel Society (ISGS) was established in 2003 as an international, interdisciplinary, not-for-profit organization whose primary purpose and objective is to promote the advancement of sol-gel science and technology. ISGS's aims are both to represent the particular needs and aspirations of the international sol-gel community and to support this sol-gel community. Prof. Plinio Innocenzi has a great contribution to ISGS as a member of founders and the first board of directors of the society. He is not only a world famous and leading scientist in the field, but also collecting respect from students and the society as a fantastic teacher. The readers can enjoy a world of sol to gel transition and understand how the ingredient chemicals change into functional materials.

ISGS convenes the biannual International Sol-Gel Conference in many parts of the world. The XIV edition of this international conference was held in Liège, Belgium, in 2017. The next one will be held in St. Petersburg, Russia, in autumn 2019. These conferences play an important role to educate, federate, and disseminate scientific knowledge to people working in related fields. To initiate young researchers and engineers into the sol-gel field, a sol-gel summer school is also operated by ISGS every 2 years in addition to the International Sol-Gel Conferences. "The Sol-to-Gel Transition" is always the central topic during these conference series and schools. This book helps readers to participate such fundamental but highly scientific discussions. After reading this book, ISGS recommends to read more extensive book series "Advances in Sol-Gel Derived Materials and

Technologies" and our official scientific journal *Journal of Sol-Gel Science and Technology*. These are also published by Springer. ISGS is very proud of having this productive relationship with Springer.

I wish you a very pleasant and educative reading!

Masahide Takahashi
President of the International Sol-Gel Society
http://www.isgs.org

Preface

The sol-gel transition is a chemical-physical process commonly observed in several systems, from organic polymers to colloids and oxides. The process, however, is not a thermodynamic event such as the glass transition, and this makes its exact determination quite difficult, especially in complex systems. Even if it is a well-studied phenomenon, defining and measuring the transition between the sol and the gel state remains a challenging task.

Many experimental and theoretical works have been dedicated to matching models with empirical results; the transition has been described in terms of random branching (classic approach) or percolation, and several sets of experiments have been specifically designed for measuring the gelation time. The intricate structure of a gel makes this evaluation not easy, especially in inorganic systems where the chemistry is particularly difficult to handle. Several books and articles have been devoted to describe many specific aspects of sol-gel inorganic chemistry, but it is still difficult to find a general treatment of the sol to gel transition. This short book has the purpose of giving a brief overview of the process in inorganic and hybrid systems. It has not the ambition to be comprehensive but rather to introduce the subject simply and descriptively, presenting some of the main theories and analytical methods which have been developed so far. The description of the process is quite general and is not addressed to specialists but rather to students and researchers who are new to the field and want to know more about the sol-gel transition in inorganic systems. Because the intention here is to provide a general overview, it has not been possible to cover all the possible topics and sub-topics, but the most important have been highlighted, and more detailed information can be found in the references.

There are several excellent reviews and handbooks dedicated to inorganic sol-gel chemistry, but it is seldom possible to find a text which is dedicated to summarize and analyze the process. An obvious limitation is that sol-gel chemistry very much depends on the peculiar conditions of synthesis, and even a small change can cause a large difference in the process. This strong dependence on the synthesis reduces the possibility to give general rules; on the other hand, the current trends in materials science are focusing the research mostly on applications, and fewer studies are devoted to understand the basic science that underlies them. Most of the examples in the book are taken from silica. There are several specific reasons for this: one is that the complete set of data is available for this oxide, and the second is that the "polymer-like" chemistry of silica makes it the most interesting of all oxides to study while also providing a more general description of the process.

The present volume is the second revised edition of the book. I have decided to add some more examples and a more comprehensive discussion of the basic sol-gel chemistry which is important to understand the process. Some errors in the text have also been corrected.

The book is now organized in seven chapters. Chapter 1 is an introduction to the concepts of sol, gel, and sol-gel transition, which are critically discussed and presented.

Chapter 2 is, instead, dedicated to the precursors of the sol-gel process. In Chap. 3, the principles of inorganic sol-gel chemistry necessary for understanding the sol to gel transition are described.

In Chap. 4, the main models used for describing the sol to gel transition are discussed, in particular, the so-called classic models and percolation theory. In contrast to other models that have been proposed, these two have been tested with experiments, and a critical discussion is therefore possible. Chapter 5 is dedicated to describe the molecular species which form at the beginning of the process and to correlate their presence with the gelation. This is a critical point to understand how the synthesis parameters affect the sol structure and, in turn, the sol-gel transition.

Chapter 6 introduces the main techniques to measure the sol to gel transition, while Chap. 7 discusses the effect of the transition on the local environment within the pores and of how to measure the sol-gel transition in fast evaporating systems.

I have made an effort, also in this revised version, to keep the common thread of the book well focused on providing a sound introduction to the key fundamentals of the sol-gel transition in inorganic systems, without the ambition to be comprehensive, or discussing all the details. Equipped with the knowledge gained herein, the reader will be prepared to explore and assess the more advanced literature on his or her own. The sol to gel transition is a basic subject that sometimes is overlooked, but a good house is always built on a solid foundation.

Sassari, Italy   Plinio Innocenzi

Contents

1 A Sol and a Gel, What Are They? 1
References... 6

2 The Precursors of the Sol-Gel Process 7
2.1 From a Precursor to a Sol, What Kind of Precursor?............. 7
2.2 Silicon Alkoxides.. 8
2.3 Exchange Reactions of Alkoxides with Alcohols 10
2.4 Organically Modified Silicon Alkoxides 10
    2.4.1 Organic Groups as Network Modifiers 12
    2.4.2 Hybrid Precursors as Coupling Agents and Surface
         Modifiers 13
    2.4.3 Alkoxides Containing Polymerizable Organic Groups 14
    2.4.4 Silsesquioxanes 14
    2.4.5 Bridged Polysilsesquioxanes 15
2.5 Transition Metal Alkoxides 15
2.6 Not Only Alkoxides: Water Glass 17
References... 18

3 From a Sol to a Gel 21
3.1 From a Sol to a Gel: Formation of a Reactive Monomer
    (Partial Hydrolysis) 21
3.2 The Point of Zero Charge.................................... 23
    3.2.1 The PZC and the Gelation Time 24
3.3 From the Precursor to a Sol, the Transition Metals 24
3.4 From a Sol to a Gel: Condensation 26
3.5 Acid-Catalyzed Hydrolysis and Condensation 26
3.6 Basic Catalyzed Hydrolysis and Condensation 27
3.7 How Much Water?... 28
3.8 Hydrolysis vs Condensation 29
3.9 Formation of the Sol 31

3.10 The Effect of Container Size on Gelation Time.   34
3.11 What About Gravity?.   35
3.12 A General View   36
References.   37

4   Sol to Gel Transition: The Models.   39
4.1   The Classic Statistical Model (Mean-Field Theory)   39
4.2   The Percolation Theory   43
      4.2.1   Percolation: What Is it?   43
      4.2.2   Site Percolation   44
      4.2.3   Bond Percolation   47
      4.2.4   Scaling and Universality   48
4.3   Percolation and sol to Gel Transition.   50
References.   53

5   From Silicate Oligomers to Gelation.   55
5.1   Failure of the Classic Model   55
5.2   Silica Cyclization.   59
5.3   The Oligomerization Pathway   62
References.   65

6   Measuring the Sol to Gel Transition   67
6.1   Measuring the Gel Point and the Gel Time   67
6.2   Measuring the Gel Point Through the Rheological Properties   68
      6.2.1   Viscosity.   68
      6.2.2   Viscoelastic Experiments for Determination
              of the Gel Point   73
      6.2.3   The Rheology of Sol-Gel Transition   80
6.3   Determination of Gel Point Through Dynamic Light Scattering. . . .   80
References.   83

7   Probing the Sol to Gel Transition in the Gel Structure   85
7.1   Diffusion During Gelation.   85
7.2   Microviscosity   87
7.3   Probing the Sol to Gel Transition in Fast Evaporating Sols   89
References.   93

Conclusion.   97

Index.   99

Chapter 1
A Sol and a Gel, What Are They?

Giving a clear and widely accepted definition in chemistry and physics is always a pretty difficult task, especially because the frontier of science is continuously moving. Time is a good friend for science because knowledge accumulates and allows getting a wider and deeper understanding of nature; however, it is also more complicated organizing what are sometimes only fragments of a big puzzle. Definitions also have a historical development, and they are modified according to the new knowledge or simply with a change of perspective. This short book is dedicated to the sol to gel transition in oxides, and therefore the first mandatory step is a clear understanding of what a gel and a sol are.

We can start from the general definition of a *sol* given by IUPAC (International Union of Pure and Applied Chemistry): "A fluid colloidal system of two or more components, e.g. a protein sol, a gold sol, an emulsion, a surfactant solution above the critical micelle concentration" [1]. This is quite general but is not self-consistent, because it needs another definition which is that one of a fluid colloidal system.

The IUPAC definition of **colloidal** is "…a state of subdivision, implying that the molecules or polymolecular particles dispersed in a medium have at least in one direction a dimension roughly between 1 nm and 1 μm, or that in a system discontinuities are found at distances of that order" [2].

The things are a little clearer, so to have a *sol* we need at least two components, one should be a fluid while the other components (one or more) are characterized by the property to have dimensions smaller than 1 μm. The not fluid part which forms a *sol* is therefore defined not by its nature but from the dimension; this gives room to include in the term "sol" a great variety of different systems.

We have, however, to pay attention because the definition of medium or better a phase in a colloidal is so general that several systems can be formed combining different media. A dispersion of solid colloidal particles in a gas forms a *solid aerosol* such as *smoke* (solid particles dispersed in a gas), while a dispersion of liquid droplets in a gas is a *liquid aerosol*, such as *fog* (water droplets suspended in air). Colloidal droplets dispersed in a liquid form an *emulsion* whose typical example is

© The Author(s), under exclusive license to Springer Nature Switzerland AG 2019
P. Innocenzi, *The Sol-to-Gel Transition*, SpringerBriefs in Materials,
https://doi.org/10.1007/978-3-030-20030-5_1

milk. If we use the IUPAC definition: *"A fluid colloidal system in which liquid drop-lets and/or liquid crystals are dispersed in a liquid. The droplets often exceed the usual limits for colloids in size"* [1], also the *emulsions* should be considered a *sol*, but in many cases the dispersing phase is formed by liquid droplets exceeding the limit of 1 μm which defines a colloid.

Maybe it is better to use a definition without including emulsions which generate some confusion. The following one is the definition of a *sol* from the Encyclopædia Britannica [3]: "Sol in physical chemistry is a colloid (aggregate of very fine parti-cles dispersed in a continuous medium) in which the particles are solid and the dispersion medium is fluid." This means that we could restrict the definition of sol to a binary system of a fluid and dispersed "solid" particles of dimensions lower than 1 μm. The particles in a *sol* do not also need to possess any particular symme-try; the shape and dimensions in the three axes are not precisely defined beside the limitation to be within 1 nm and 1 μm. No particular limitations are also given to the particle structure, and they can be amorphous, crystalline, porous, or dense.

But, what about the micelles, should they be considered a solid? Micelles form in a liquid containing a surfactant when a critical concentration is reached, but they are a supramolecular aggregate and not properly a solid. A liquid containing micelles as the dispersed phase is, therefore, a *sol* or not? It depends on the definition we use; what is clear is that we can fix some boundaries, while giving a fully comprehensive description of *sol* is still a difficult task.

In conclusion, we can state that a *sol* represents, therefore, a particular case of a colloidal system given by the combination of a liquid as dispersive medium and a solid as the dispersed phase, with the limit of dimension which has been just defined (smaller than 1 μm). Why a critical dimension of 1 μm has been defined, is there any special reason for this? The answer is that if the dimension of the particles remains below this value, the equilibrium of the system is governed by Brownian motion. Beyond this dimension (which has to be taken of course as a theoretical average value), the gravity force becomes predominant over short-range forces and pushes the particles to sediment. Putting the limit of 1 μm, we implicitly assume that the *sol* should be stable, and it is not necessary to specify this property in the definition.

If we look back to the IUPAC definition, we can notice that the particles can be either molecular or macromolecular, which excludes the case of solid nanoparticles, metallic or semiconductors, for instance. Gold nanoparticles appear instead in the definition of sol which is a colloidal... this is just to underline how much giving consistent and coherent definitions in chemistry could be a difficult task.

To make the things even more clear, we have to stress that the terms *solution* and *sol* refer to something quite different; in a solution, in fact, the solvent and the solute form only one phase, while in a colloid such as sol, the phases must be at least two, a dispersed phase of suspended particles and a continuous phase, the liquid, which is the medium of suspension. It is all? Almost... in fact, if the suspended particles precipitate in a short time, our definition of a colloid is not so useful; this means therefore that the temporal stability of the system is also important. How long should a colloid be stable? It is better not to give a well-defined span of time and is enough to state that it should take a "very long time" before observing precipitation of the solid phase. This should be enough to be sure that the gravitational forces on the particles are not effective while short-range interactions, such as van der Waals and surface charges, predominate as we have seen.

Because the subject of this book is the sol to gel transition, some other questions arise. The first one is due to the nature of the solid particles, which actually could be a polymer or a molecule which reacts once dissolved in a liquid forming larger aggregates. If these aggregates do not exceed 1 μm, we still have a sol, but… most likely we cannot go back… and once the reaction has started, the colloid becomes an *irreversible* system [4]. This case is not, however, so general, and in some systems, reversibility is also possible, in particular when secondary bonds, such as hydrogen and van der Waals, are involved. This type of gel is indicated as a *physical gel*, where entanglements and secondary bonds contribute to gelation to distinguish from *chemical gels* where irreversible covalent bonding is instead generally involved. Physical gels, such as aqueous gelatine, undergo a thermo-reversible transition from gel to sol (*reversible transition from liquid-like to solid-like state*). This response is found in a broad range of macromolecular systems, biological as well as synthetic polymers (*polymer solutions*). Gelation of organic polymers is a widely studied subject, but the chemistry and physics can be quite different from the inorganic sol-gel process even if the general principles are similar.

If the reactions (polymerization and/or polycondensation) do not stop when the colloidal dimension has been reached, the further growth forms a continuous solid phase which expands throughout all the liquid. This consideration is very close to the IUPAC definition of a *gel*: "Non-fluid colloidal network or polymer network that is expanded throughout its whole volume by a fluid" [5]. The liquid is, therefore, entrapped within a solid network and is responsible for the expansion of the solid phase; intuitively, therefore, any removal of the liquid through a drying process produces shrinkage of the system. The concept of continuity of the phases in a gel is also very important and has been well described by Jeffrey Brinker and George Scherer: "Continuity means that one could travel through the solid phase from one side of its sample to the other without having to enter the liquid; conversely, one could make the same trip entirely within the liquid phase" [6]. We can image a trip of two small nanocapsules, the first one travels within a liquid tunnel along with all the gels and the other one in a kind of highway formed by the solid part of the same gel. Whenever complicate will be the route, and even if by chance they could come very close (even few nanometers), they would never get in touch.

What is making the definition of gel so tricky is exactly this coexistence of two phases, a continuous solid one (the spanning macromolecule) and the liquid component (the residual sol). This difficulty to describe a gel [7] has been well-defined by Hench [8]: "A gel, for instance, has been defined as a 'two-component system of a semi-solid nature rich in liquid' and no one is likely to entertain illusion about the rigor of such a definition."

In the definition of gel, the liquid phase can be any fluid without a particular restriction. If the solid phase is a hydrophilic polymer and the liquid is water, this is a very peculiar type of gel which is indicated as a *hydrogel*.

There is another property of gels which is important to underline; they exhibit a mechanical resistance, whenever small, and the formation of a gel is associated with the resistance to shear stress and elastic deformation. This is quite helpful for defining a gel, and monitoring the change of mechanical properties is one of the best ways of observing the transition from a sol to a gel. The development of an elastic response is what distinguishes a gel from a sol, and this difference can be also used to give a more precise definition of gel (Encyclopædia Britannica): "Coherent mass consisting of a liquid in which particles too small to be seen in an ordinary optical microscope are either dispersed or arranged in a fine network throughout the mass. A gel may be notably elastic and jellylike (as gelatin or fruit jelly), or quite solid and rigid (as silica gel). Gels are colloids in which the liquid medium has become viscous enough to behave more or less as a solid" [9].

Another observation is that a gel is a disordered state, and in the case of silica, the gel is an intermediate state to obtain a glass. After drying (removal of residual solvent) and firing, the final product is a high-purity silica glass which cannot be distinguished by the structure from a glass of similar composition prepared from high temperatures [10].

As we have seen, the family of gel materials is very large, including organic polymers, oxides, and organic-inorganic hybrids. We have now to restrict our discussion to the case which is known as a sol-gel process and is generally associated with oxides and hybrids. This process has some peculiarities and, at least at the very beginning, has been used for obtaining ceramic materials from a low-temperature route.

We need, however, to restrict the boundary of the sol-gel process; it means that another definition needs to be introduced, and again we use the IUPAC one: "Process through which a network is formed from solution by a progressive change of liquid precursor(s) into a sol, to a gel, and in most cases finally to a dry network" [11]. This definition also has a note: "An inorganic polymer, e.g., silica gel, or an organic–inorganic hybrid can be prepared by sol-gel processing." Following IUPAC the sol-gel process involves, therefore, the transformation from a colloidal to a gel system; at the end of the process, when a gel is finally obtained, the removal of the liquid through drying gives a solid material.

Almost all the definitions we need have been introduced; we are almost at an end, but it is still necessary to define the *gel point* and finally the *sol-gel transition* which is the main subject of this book. The gel point is the: "Point of incipient network formation in process forming a chemical or physical polymer network" [12]. The notes which accompany this definition are very important; the first one is: "In both network-forming polymerization and the crosslinking of polymer chains, the gel point is expressed as an extent of chemical reaction." This means that the percentage of bond conversion is an important parameter for modeling the process and making a prevision about gelation.

The other interesting note is: "The gel point is often detected using rheological methods. Different methods can give different gel points because viscosity is tending to infinity at the gel point and a unique value cannot be measured directly." The development of a mechanical resistance [13], in particular to shear stress, is a property that can be used not only to define a gel but also to measure the gel point.

Now we have reached the very last definition, we know what is a *sol*, a *gel*, a *sol-gel process*, a *gel point*, and it is only remaining to state, keeping in mind the previous paragraphs, what is a *sol-gel transition*: *the transition of a sol to a gel at the gel point* [14].

All the definitions have been introduced, and the main concepts we need should be there, but let's make a little confusion again. We have previously underlined that a gel is a disordered material [15], such as glass, and a sol can go through a gel transition and then a glass transition. There are, however, some systems of interacting particles whose increase of density, for instance, in colloidal suspensions, can cause a "non non-equilibrium transition from a fluid-like to a solid-like state, characterized solely by the sudden arrest of their dynamics" [16, 17]. These transitions are generally identified as *crowding* or *jamming phenomena* which freeze the disordered fluid-like state of the system. The phenomenon is quite general and allows introducing a new perspective on describing transitions from a sol to a solid material. The concept of jamming is no alternative to sol-gel transition but is a possible model for describing changes in colloidal systems and unifying the glass formation, gelation, and colloidal aggregation processes within one chemical-physical phenomenon [18].

To end this chapter, it remains to introduce the other two terms which are often encountered in the scientific literature regarding sol-gel processing of inorganic materials, *xerogel* and *aerogel*. They are a little confusing. In fact, they are not at all a gel, but because they are derived from a gel, they have been defined using a word which contains the suffix gel. The first one, *xerogel*, is sometimes used to identify the material obtained upon drying of a gel. In the case of a sol-gel film, for instance, what is obtained after the deposition is a gel layer which contains a liquid phase, mainly water and residual solvent; after drying at temperatures low enough to remove the solvent but avoiding the full densification, the dried material which is produced is defined as *xerogel* (dense gel). If the liquid phase in a bulk gel is replaced by a gas, a material with extremely high porosity (up to 99.8%) and air as the main component is obtained. This material is defined as an *aerogel* and can be fabricated from a gel upon extraction of the liquid phase via supercritical drying.

Sometimes in the literature are also used other expressions, such as a *wet gel* or *dried gel*; again these definitions should be used with caution because a gel is by definition wet, and a dried gel without the liquid phase is not a gel.

But at the end of the days, what kind of material is a gel? Is it a solid, semisolid? It is "almost" a solid because under stress it shows a mechanical resistance and an elastic response; however, this response differs from that one of a solid which does not exhibit the same extent of structural deformation under small applied stresses. Liquids and gases on turn show a continuous deformation under the effect of a small external force, while gels do not. So a gel should be something between, and it is very helpful to close this long discussion introducing another concept which is that one of *soft matter*. Any material which is easily deformed by thermal fluctuations and small external forces can be considered as *soft matter*. An inorganic or hybrid gel easily falls in this large family of fascinating materials. Defining more in detail what is a soft material will take much more space and is beyond our purpose, but we keep this idea in mind as a different way of thinking about gels.

In the next chapters, we will mostly concentrate on the sol to gel conversion, but we need of course to have cleared in mind on how a sol forms and condenses up to the gel point. This is the subject of the second chapter.

References

1. PAC, 1972, 31, 577. Manual of symbols and terminology for physicochemical quantities and units, Appendix II: definitions, terminology and symbols in colloid and surface chemistry, p 606
2. PAC, 1972, 31, 577. Manual of symbols and terminology for physicochemical quantities and units, Appendix II: definitions, terminology and symbols in colloid and surface chemistry, p 605
3. http://global.britannica.com/science/sol-colloid
4. Matsoukas T (2015) Statistical thermodynamics of irreversible aggregation: the sol-gel transition. Sci Rep 5:8855
5. PAC, 2007, 79, 1801. Definitions of terms relating to the structure and processing of sols, gels, networks, and inorganic-organic hybrid materials (IUPAC recommendations 2007), p 1806
6. Brinker J, Scherer G (1990) Sol-gel science. Academic Press, Boston
7. Almdal K, Dyre J, Hvidt S, Kramer O (1993) Towards a phenomenological definition of the term 'Gel'. Polym Gels Netw 1:5–17
8. Henish HK (1970) Crystal growth in gels. The Penn State University Press, University Park
9. Encyclopædia Britannica. www.britannica.com/science/gel
10. James PF (1988) The gel to glass transition: chemical and microstructural evolution. J Non-Cryst Solids 100:93–114
11. PAC, 2007, 79, 1801. Definitions of terms relating to the structure and processing of sols, gels, networks, and inorganic-organic hybrid materials (IUPAC recommendations 2007), p 1825
12. PAC, 2007, 79, 1801. Definitions of terms relating to the structure and processing of sols, gels, networks, and inorganic-organic hybrid materials (IUPAC recommendations 2007), p 1809
13. Grant MC, Russel WB (1993) Volume-fraction dependence of elastic moduli and transition temperatures for colloidal silica gels. Phys Rev E 47:2606–2614
14. PAC, 2007, 79, 1801. Definitions of terms relating to the structure and processing of sols, gels, networks, and inorganic-organic hybrid materials (IUPAC recommendations 2007), p 1826
15. Innocenzi P, Malfatti L, Kidchob T, Falcaro P (2009) Order–disorder in self-assembled meso-structured silica films: a concepts review. Chem Mater 21:2555–2564
16. Biroli G (1997) Jamming: a new kind of phase transition? Nat Phys 3:222–223
17. Trappe V, Prasad V, Cipelletti L, Segre PN, Weitz DA (2001) Jamming phase diagram for attractive particles. Nature 411:772–775
18. Sciortino F, Buldyrev SV, De Michele C, Foffi G, Ghofraniha N, La Nave E, Moreno A, Mossa S, Saika-Voivod I, Tartaglia P, Zaccarelli E (2005) Routes to colloidal gel formation. Comp Phys Comm 169:166–171

Chapter 2
The Precursors of the Sol-Gel Process

Abstract Sol-gel processing is a highly versatile method and allows obtaining a large variety of materials of different composition such as oxides, mixed oxides, and hybrid organic-inorganic materials. The chemistry of the process largely depends on the choice of precursors. Inorganic salts, metal alkoxides, and organosilanes are some of the most common components used for sol-gel processing, and so many synthesis routes have been developed for the design of materials with highly tailored properties. This chapter is a brief overview of the main sol-gel precursors.

Keywords Alkoxides · Organosilanes · Surface modifiers · Silsesquioxanes · Bridged silsesquioxanes · Water glass

2.1 From a Precursor to a Sol, What Kind of Precursor?

The IUPAC definition of the sol-gel process, which we have already introduced in the previous chapter: "… *process through which a network is formed from solution by a progressive change of liquid precursor(s) into a sol, to a gel…*" [1], indicates that the formation of an inorganic chemical gel is the result of a gradual transformation. Even if the process is continuous, different stages [2] can be defined with the caution that a clear identification of the single steps is hampered by the overlapping of several competing processes, such as hydrolysis and condensation.

The first stage of the process is the formation of a sol from a precursor; the second is the development of a gel through the sol to gel transition (Fig. 2.1). This transformation is due to the reactions of hydrolysis and condensation and the progressive conversion of the reactive sites into bridging bonds.

The scheme looks a little bit simplistic but is quite effective to resume the different stages that are involved in the process, the precursor, the sol, the transition from the sol to the gel, and the gel phase keeping in mind that hydrolysis and condensation cannot be separated and are interdependent. The final material, a dense oxide or hybrid organic-inorganic, is obtained only after removal of the residual solvent (*drying*) and full condensation of the network (*firing*) [3, 4].

© The Author(s), under exclusive license to Springer Nature Switzerland AG 2019
P. Innocenzi, *The Sol-to-Gel Transition*, SpringerBriefs in Materials,
https://doi.org/10.1007/978-3-030-20030-5_2

Fig. 2.1 Scheme of the
stages involved in a sol to
gel transition in inorganic
or hybrid systems

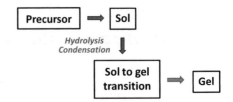

Let's now have a look to the precursors; in general, they can be either an *inorganic metal salt* (chloride, nitrate, sulfate, etc.) or a *metal alkoxide* [5]. This last one is part of the bigger family of *metalorganic compounds* [5, 6], and they are the most widely used precursors, because they react easily with water and are available for many metals. Metal alkoxides have a peculiar structure which is formed by a metal atom bonded to one or more alkoxy groups by an intermediate oxygen atom.

There is an important difference between silicon and transition metal alkoxides; the chemistry of the silicon alkoxides is relatively easier to handle in comparison to transition metal alkoxides (such as Ti, Al, Zr), which are much more reactive. In the case of silicon alkoxides, it is necessary a catalyst to start the reactions, while for the transition metals, on the contrary, chelating agents have to be used to avoid a fast reaction with water and immediate precipitation.

2.2   Silicon Alkoxides

The most popular family of precursors for sol-gel silica processing is composed by silicon alkoxides (alkoxysilanes). They are characterized by the strong covalent Si-O bonding and are hydrophobic and immiscible with water.

Tetraethyl orthosilicate (TEOS), $Si(OC_2H_5)_4$, is the first alkoxide of the series, followed by tetramethyl orthosilicate (TMOS), $Si(OCH_3)_4$, which is, however, less safe to handle and hydrolyzes faster than TEOS (Fig. 2.2).

The hydrolysis of TMOS is, in fact, around six times faster: in general, a lower hydrolysis rate is associated with an increase of the organic group size in the silicon alkoxide (Fig. 2.3). The properties of the silicon alkoxides change according to the dimension of the alkoxy; larger groups produce an increase in molecular weight, viscosity, and boiling point and a decrease in density of the alkoxides (Fig. 2.4).

As a rule of thumb, a larger size of the alkoxy group is associated with a lower hydrolysis rate due to the steric hindrance. The reactivity follows the sequence (2.1), with tetramethyl orthosilicate the most reactive alkoxide:

$$Si(OCH_3)_4 > Si(OC_2H_5)_4 > Si(n\text{-}OC_3H_7)_4 > Si(n\text{-}OC_4H_9)_4$$

| tetramethyl orthosilicate | > | tetraethyl orthosilicate | > | tetra-n-propylorthosilicate | > | tetrabutyl orthosilicate | (2.1) |

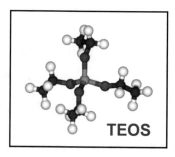 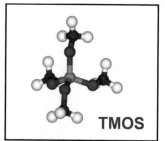

Fig. 2.2 Stick and ball models of tetraethyl orthosilicate (TEOS), *left*, and tetramethyl orthosilicate (TMOS), *right*

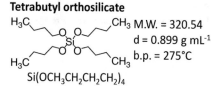

Tetramethyl orthosilicate

M.W. = 152.22
d = 1.03 g cm^{-3}
b.p. = 121°C

$Si(OCH_3)_4$

Tetraethyl orthosilicate

M.W. = 208.33
d = 0.94 g cm^{-3}
b.p. = 168°C

$Si(OC_2H_5)_4$

Tetrabutyl orthosilicate

M.W. = 320.54
d = 0.899 g mL^{-1}
b.p. = 275°C

$Si(OCH_3CH_2CH_2CH_2)_4$

Tetrapropyl orthosilicate

M.W. = 264.43
d = 0.916 g mL^{-1}
b.p. = 224°C

$Si(OCH_3CH_2CH_2O)_4$

Fig. 2.3 The most common silicon alkoxides and their properties

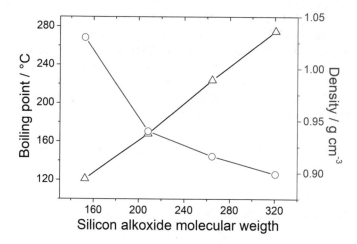

Fig. 2.4 Change of boiling point (°C), left y-axis, and density (g cm^{-3}), right y-axis, as a function of molecular weight in silicon alkoxides

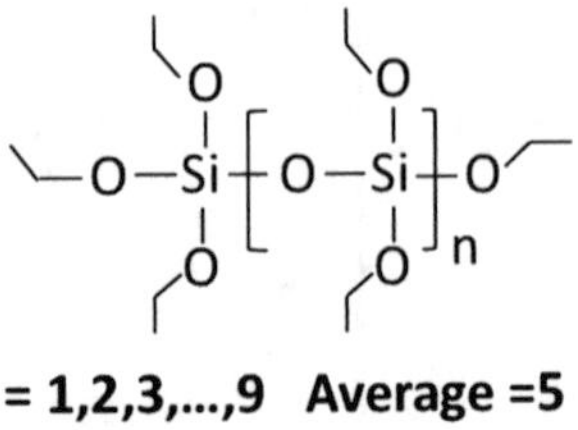

Fig. 2.5 Ethyl silica 40
average formula

n = 1,2,3,...,9　**Average =5**

Silicon alkoxides such as TEOS are commonly prepared via alcoholysis of the silicon chloride (2.2):

$$SiCl_4 + 4C_2H_5OH \rightarrow Si(OC_2H_5)_4 + 4HCl \tag{2.2}$$

An alternative source for the synthesis of silicate gels is a hydrolyzed and oligomerized form of ethyl silicate [7–9]. A commercial one, for instance, is ethyl silicate 40 (ES40) which is a mixture of monomers, dimers, trimers, and cyclic silanes with 40% silica by mass. ES40 can be considered a partially condensed form of TEOS, with an average of five silicon atoms per molecule (Fig. 2.5). In general, because the silica molecules are partially condensed, the solubility in aqueous solutions is lower than TEOS. The boiling point is higher (290–310 °C) than TEOS as such as the viscosity and the density. An advantage of ES40 is the lower cost with respect to silicon alkoxides.

2.3 Exchange Reactions of Alkoxides with Alcohols

An important property of metal alkoxides is the capability of activating exchange reactions with alcohols. This has to be taken into account during sol-gel processing; in fact, the presence of alkoxides with mixed ligands affects the hydrolysis and condensation reactions because they have a different reactivity and solubility. In general, mixing a metal alkoxide with different alcohol produces almost immediately an exchange reaction with the formation of mixed ligands [10]. An example is the transesterification reaction of tetramethyl orthosilicate (TMOS) when dissolved in ethanol; the equilibrium reaction with ethanol becomes (2.3):

$$Si(OCH_3)_4 + C_2H_5OH \rightleftharpoons Si(OCH_3)_4(OC_2H_5) + CH_3OH \tag{2.3}$$

2.4 Organically Modified Silicon Alkoxides

Silicon is also able to form a strong and hydrolytically stable bond with carbon; it is, therefore, possible the synthesis of a family of organo-substituted derivatives (organically modified alkoxides) with different functionalities. This is a quite large family of precursors whose properties depend on the substitute organic groups which are covalently attached to the alkoxysilyl groups via the Si-C bonds [11]. These groups are characterized by different geometry, length, rigidity, and functionality.

The organically modified alkoxides can be mainly divided into three groups, on the basis of the role played by organic groups. The final product is a hybrid organic-inorganic material characterized by the presence in the backbone of carbon atoms covalently bonded to silicon [12]. The materials have in general intermediate properties between inorganic oxides and organic polymers [13].

These organosilanes can form a hybrid material by themselves, but in general, they are co-reacted with another alkoxide, such as TEOS, to obtain the final organic-inorganic product. This requires careful control of the kinetics of the reaction to obtain a homogeneous material without phase separation [14].

The true nature of hybrid organic-inorganic materials synthesized via sol-gel processing is, however, difficult to define with clarity. In general, they are characterized by a direct covalent chemical bond which connects the inorganic and organic species. Incorporation of organic molecules into an oxide matrix, which can be easily done because the material synthesis is performed at low temperatures through solution processing, gives in most of the cases the formation of composites at the molecular level [15] even if they are considered by some authors also a particular type of hybrid.

An example is the incorporation of fluorescent dyes [16], such as rhodamine 6G [17] or rhodamine B into a sol-gel matrix. The final material can be considered a composite at the molecular level because both the guest molecule and the matrix do not change the chemical-physical properties. The surrounding chemical environment can affect the optical response, but rhodamine B can be still identified as a single molecule [18]. On the other hand, if rhodamine B is modified to form a silylated dye (Fig. 2.6), it can be directly used during the sol-gel reactions to form a hybrid. In this case, the rhodamine B covalently connects to the silica network [19]; the molecule not only modifies the structure of the material, affecting, for instance, its mechanical properties but adds a function, which is the optical response.

The identification of a hybrid organic-inorganic sol-gel material is, therefore, not always so straightforward, and a clear-cut definition is difficult. Materials formed by interpenetrating organic and inorganic networks are another interesting example of hybrid composite. They can be obtained via sol-gel processing by independent polymerization of the organic and inorganic networks [20]. The possibility of forming a homogeneous composite depends on the capability of obtaining comparable rates of organic polymerization and inorganic polycondensation during the synthesis. When the organic polymerization is faster than the formation of an extended inorganic network via sol-gel reactions and vice versa, a heterogeneous structure with phase separation is observed.

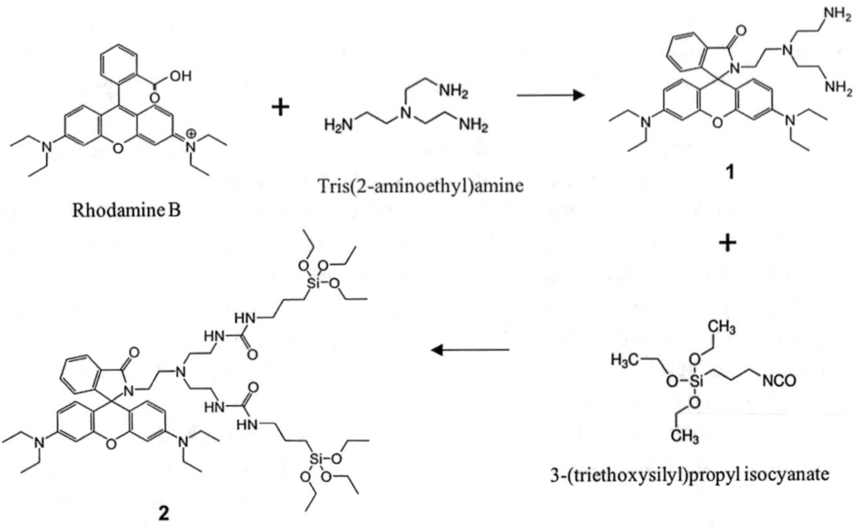

Fig. 2.6 Synthesis of a rhodamine B derivative as a precursor of hybrid functional material. The reaction of rhodamine B with tris(2-aminoethyl)amine gives the intermediate **1**, and further reaction with 3-(triethoxysilyl)propyl isocyanate allows forming the hybrid precursor **2**

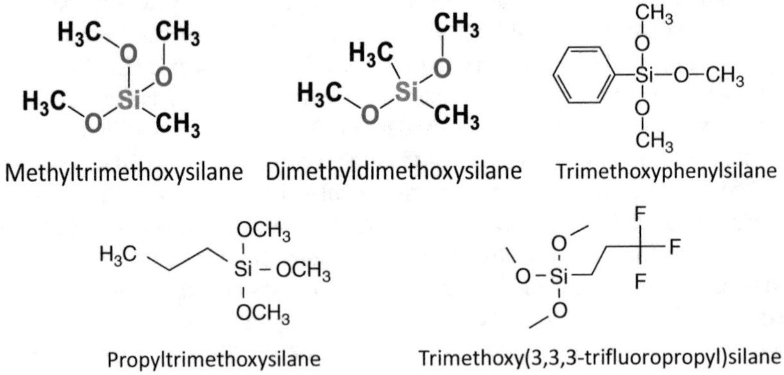

Fig. 2.7 Silicon derivatives containing organic groups as modifiers

2.4.1   Organic Groups as Network Modifiers

This group of alkoxides contains some of the most popular precursors for the preparation of hybrid materials, such as methyltrimethoxysilane (MTES, CH_3-$SiO(CH_3)_3$), dimethyldimethoxysilane, and phenyltrimethoxysilane (Fig. 2.7). The functional organic group can also contain fluorine atoms, such in the case of trimethoxy(3,3,3-trifluoropropyl)silane.

These alkoxides are characterized by a general formula R'Si(OR)$_3$, but also bi (R'$_2$Si(OR)$_2$) or three (R'$_3$Si-OR) substitute alkoxides can be employed. It is now commercially available a very wide range of precursors with different functional groups such as amine, isocyanate, thiol, amide, polyether, and so on.

The covalently bonded organic groups modify the structure and the property of the hybrid materials; changing the amount and type of the functional group will produce a direct modification of the properties, such as the wettability, the mechanical resistance, and the optical properties. The reactivity of the modified alkoxides would also be affected by the functional groups.

2.4.2 Hybrid Precursors as Coupling Agents and Surface Modifiers

Another peculiar class of hybrid precursors is formed by alkoxides that are commonly used as coupling agents or surface modifiers (Fig. 2.8). Some examples are (3-aminopropyl)triethoxysilane (APTES, H$_2$N(CH)$_3$Si(OC$_2$H$_5$)$_3$), [3-(2-aminoethylamino) propyl]trimethoxysilane, (AEPTES, (CH$_3$O)$_3$Si(CH$_2$)$_3$NHCH$_2$CH$_2$NH$_2$), and (3-mercaptopropyl)trimethoxysilane (HS(CH$_2$)$_3$Si(OCH$_3$)$_3$).

APTES is somehow a special type of precursor, because of the presence of a primary amine which can also easily react with many other organic species, such as epoxides, to form hybrid materials of more complex structure [21]. APTES because of its properties as also one of the most popular coupling agents. For instance, can react with surface silanols via its alkoxy function, while the amine can react with organic resins such as epoxy. Another peculiarity of APTES is that the amine can promote the polycondensation via a self-catalytic effect produced by an increase of the sol pH.

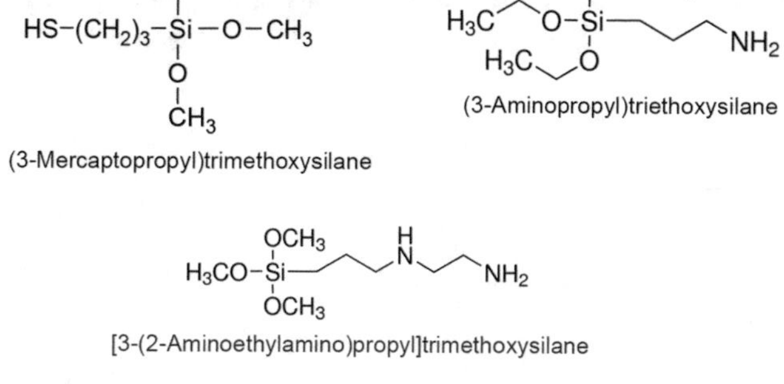

Fig. 2.8 Organosilanes used as coupling agent and surface modifier

2.4.3 Alkoxides Containing Polymerizable Organic Groups

In this group, we can find different types of alkoxides which contain polymerizable functions in the organic group, such as epoxy (3-glycidoxypropyltrimethoxysilane, GPTMS) [22, 23], vinyl (vinyltrimethoxysilane, VTMS), or methacrylate (3-metha crylodoxypropyltrimethoxysilane (MPTMS)) (Fig. 2.9). A typical example is GPTMS which has an epoxy ring whose opening allows the formation, under controlled conditions, of a poly(ethylene oxide) chain. The organic polymerization is simultaneously achieved with the formation of the inorganic network; in general, higher will be the condensation of the silica network and shorter will be the organic chains because of the smaller room for growth within the gel structure and vice versa. The simultaneous formation of an inorganic and organic network is a competitive process. The hybrids which result from this class of alkoxides may contain, therefore, an organic polymer whose extent depends on the synthesis conditions.

This class of organosilanes allows producing a large variety of hybrid materials. It is, in fact, possible to form mixed oxides networks through sol-gel reaction with another metal alkoxide but also obtaining an extensive polymerization of the organic side by reaction with organic polymer precursors. An example is the reaction of 3-methacryloxypropyltrimethoxysilane with the methacrylate monomers. It is also possible to form true hybrid composites, which are generally indicated as interpenetrated hybrids, by forming inorganic and organic networks which do not have, however, any direct covalent bond.

2.4.4 Silsesquioxanes

A particular class of hybrid precursors is formed by cage-like structured organosilicon molecules which have Si-O-Si bonds and silicon atoms at the tetrahedral vertices; silsesquioxanes may also have a polymeric structure with a ladder-like repeating unit [24] or random or open-cage structures [25]. Silsesquioxanes have a general formula $(RSiO_{1.5})_n$ with the substituent R = H, alkyl, aryl, or alkoxy. The composition explains the name because every silicon is linked in average to one and a half

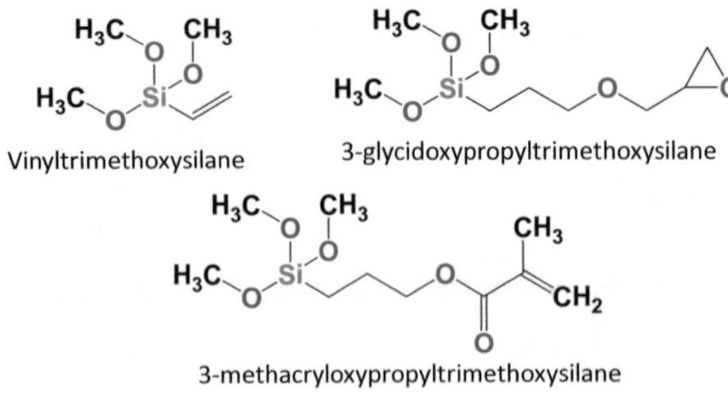

Fig. 2.9 Silicon derivatives modified with polymerizable organic functional groups

Vinyltrimethoxysilane

3-glycidoxypropyltrimethoxysilane

3-methacryloxypropyltrimethoxysilane

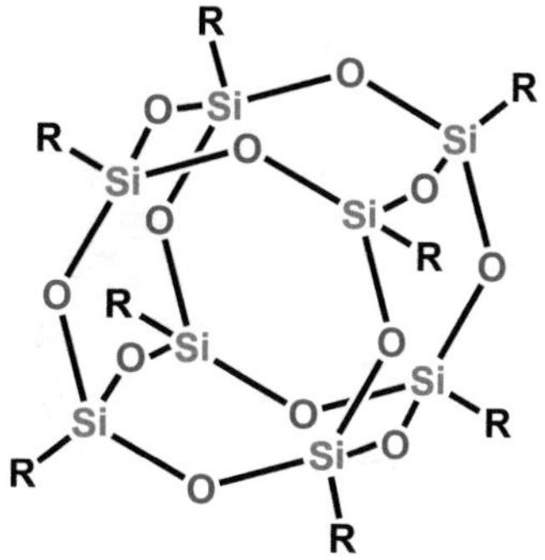

Fig. 2.10 The structure of
a polyhedral oligomeric
silsesquioxane (POSS)

(*sesqui*) oxygen atoms and to one hydrocarbon group (*ane*). The functional groups give the property of the silsesquioxanes and can also be hydrolyzed and condensed in the case of alkoxy, chlorosilanes, silanols, and silanolates. It is also possible to synthesize well-defined structures defined as polyhedral oligomeric silsesquioxanes (POSS) (Fig. 2.10).

2.4.5   Bridged Polysilsesquioxanes

A particular type of hybrid materials can be obtained by employing bridged silses-quioxanes molecular precursors (R'O)₃SiRSi(OR)₃ [26, 27]. They are characterized by an organic spacer which is bridging two or more silicon atoms. If the nature of the organic spacer, R, and the synthesis are carefully designed, it is possible to obtain a hybrid material with long-range structural order. The precursor has clearly a higher number of available siloxane linkages, in the case of two silicon atoms, six instead of four, which changes its reactivity in the sol-gel process.

Self-organization into a crystalline hybrid structure has been observed in bridged polysilsesquioxanes with different types of organic spacers (Fig. 2.11); lamellar crystals [28] but even helical fibers have formed during gelation of a bulk hybrid gel.

2.5   Transition Metal Alkoxides

The chemistry of transition metal alkoxides is much different from that one of sil-ica; silicon is in fact tetrahedrally coordinated to oxygen, while metals have usually an octahedral coordination [29, 30]. The tetrahedral structure of silica, because it is much more flexible than octahedras, is able to form "polymeric" structures of dif-ferent types. Nonsilicate metal alkoxides are very reactive with water; the hydroly-sis rate of a titanium alkoxide (Fig. 2.12) is generally up to 10^5 times faster than for

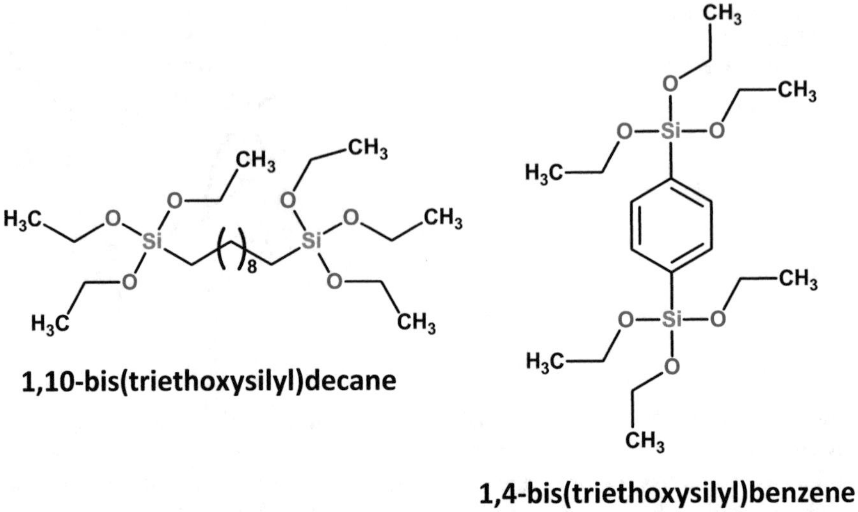

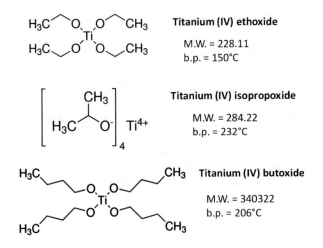

1,10-bis(triethoxysilyl)decane

1,4-bis(triethoxysilyl)benzene

Fig. 2.11 Examples of bridged polysilsesquioxanes: 1,10-bis(triethoxysilyl)decane and 1,4-bis(triethoxysilyl)benzene

Titanium (IV) ethoxide

M.W. = 228.11
b.p. = 150°C

Titanium (IV) isopropoxide

M.W. = 284.22
b.p. = 232°C

Titanium (IV) butoxide

M.W. = 340322
b.p. = 206°C

Fig. 2.12 Titanium alkoxides most commonly employed in sol-gel synthesis

the corresponding silicon alkoxide. They are salts of alcohol or acids and react as strong bases which make them extremely reactive with water. The hydrolysis and condensation reactions must be, therefore, controlled by using complexing ligands, such as acetylaceton, which inhibit condensation reactions and avoid precipitation.

Mixed oxides are also commonly prepared via sol-gel, but the relative reactivity of the alkoxides involved in the process has to be carefully controlled to avoid phase separation.

Table 2.1 Electronegativity (χ), coordination number (N), and degree of unsaturation (N − Z) for some metals alkoxides

Alkoxides	χ	N	Z	N − Z
Si(OPr^i)$_4$	1.90	4	4	0
Sn(OPr^i)$_4$	1.96	6	4	2
Ti(OPr^i)$_4$	1.54	6	4	2
Zr(OPr^i)$_4$	1.33	7	4	3
Ce(OPr^i)$_4$	1.12	8	4	4
Al(OPr^i)$_3$	1.61	6	3	3

Is it possible to know "a priori" if a metal alkoxide is more reactive than another one? This is an interesting question and the answer is yes, at least if we are satisfied to get a qualitative scale of reactivity. If for any specific atom of an alkoxide is considered the coordination number, N, and the oxidation state, Z, the difference between these two values, N − Z, which has been defined as unsaturation number, gives a good indication of the reactivity. Table 2.1 lists the electronegativity and the degree of metal unsaturation of some metal alkoxides [31]. Lower is the unsaturation number and lower will be also the reactivity. The data show that the silicon alkoxide has comparatively the lowest degree of unsaturation together with a low electrophilicity and exhibits, therefore, the lowest reactivity in the series of alkoxides.

2.6 Not Only Alkoxides: Water Glass

The sol-gel chemistry of silica is not limited to silicon alkoxide as precursors; there is another class of silicon-based compounds which is commonly employed, aqueous solutions of silicates. They are indicated as *water glass* because the solvent for the reactions is water. This marks an important difference with silicon alkoxides which are instead immiscible in water and require in general an alcohol as cosolvent for the reaction. Another difference is that in water glasses the process is initiated by a pH change while in the alkoxides by the addition of water and the catalyst.

The silica glass precursors have the general formula (2.4):

$$mSiO_2 \cdot M_2O \cdot nH_2O \tag{2.4}$$

with M the alkali metal and m the molar ratio which defines the number of silica moles per oxide metal (M_2O).

The average composition of silicate species in water glass solutions is M_2SiO_3 (with M = Na or K); the solutions of water glass are formed by mixtures of monomeric and oligomeric silicates with negatively charged non-bridging oxygen.

In the case of sodium silicate, for instance, the hydrolysis reactions are initiated by the addition of hydrochloric acid (2.5):

$$\equiv \text{Si-O-Na}^+ + \text{H}_3\text{O}^+\text{Cl}^- \rightarrow \equiv \text{Si-OH} + \text{Na}^+\text{Cl}^- \quad (2.5)$$

and the condensation by reaction of two silanols.

The stability of water glass solutions is reached only in strongly basic conditions when the anionic species repel each other. On the other hand, even if the equilibrium of water glass solutions depends on several parameters, such as temperature, concentration, pH, etc., the higher complexity chemistry makes the silicon alkoxides a much more flexible precursor for sol-gel processing.

References

1. Detailed descriptions of sol-gel inorganic materials and processes can be found in two comprehensive handbooks: 1. Handbook of sol-gel science and technology, ed. by Sumio Sakka. Kluwer Academic Publishing. 2. The sol-gel handbook, ed. by David Levy and Marcos Zayat. Wiley-VCH, 2015, and the "classic" textbook of Jeffrey Brinker and George Scherer "Sol-Gel Science", Academic Press 1990
2. Orgaz F, Rawson H (1986) Characterization of various stages of the sol-gel process. J Non-Cryst Solids 82:57–68
3. Scherer GW (1988) Aging and drying of gels. J Non-Crystal Solids 100:77–92
4. Brinker CJ (1994) Sol-gel processing of silica, the colloid chemistry of silica, chapter 18. Adv Chem 234:361–401
5. Turova NY, Turevskaya EP, Kessler VG, Yanovskaya MI (2002) The chemistry of metal alkoxides. Kluwer Academic Publishers, Dordrecht
6. Bradley DC, Mehrotra R, Rothwell I, Singh A (2001) Alkoxo and aryloxo derivatives of metals. Academic Press, San Diego
7. Cihlar J (1993) Hydrolysis and polycondensation of ethyl silicates. 2. Hydrolysis and polycondensation of ETS40 (ethyl silicate 40). Colloids Surf A Physicochem Eng Asp 70:253–268
8. Wang S, Wang DK, Jack KS, Smart S, Diniz da Costa JC (2014) Improved hydrothermal stability of silica materials prepared from ethyl silicate 40. RSC Adv 5:6092–6099
9. Wang S, Wang DK, Smart S, Diniz da Costa JC (2015) Ternary phase-separation investigation of sol-gel derived silica from ethyl silicate 40. Sci Rep 5:14560
10. Dong H, Lee M, Thomas RD, Zhang Z, Reidy RF, Mueller DW (2003) Methyltrimethoxysilane sol-gel polymerization in acidic ethanol solutions studied by ^{29}Si NMR spectroscopy. J Sol-Gel Sci Technol 28:5–14
11. Jitianu A, Britchi A, Deleanu C, Badescu V, Zaharescu M (2003) Comparative study of the sol–gel processes starting with different substituted Si-alkoxides. J Non-Cryst Solids 319:263–279
12. Schubert U, Husin N, Lorenz A (1995) Hybrid inorganic-organic materials by sol-gel processing of organofunctional metal alkoxides. Chem Mater 7:2010–2027
13. Judeinstein P, Sanchez C (1996) Hybrid organic-inorganic materials: a land of multidisciplinarity. J Mater Chem 6:511–525
14. Fyfe CA, Aroca PP (1997) A kinetic analysis of the initial stages of the sol-gel reactions of methyltriethoxysilane (MTES) and a mixed MTES/tetraethoxysilane system by high-resolution ^{29}Si NMR spectroscopy. J Phys Chem B 101:9504–9509
15. Sanchez C, Boissiere C, Cassaignon S, Chaneac C, Durupthy O, Faustini M, Grosso D, Laberty-Robert C, Nicole L, Portehault D, Ribot F, Rozes L, Sassoye C (2014) Molecular engineering of functional inorganic and hybrid materials. Chem Mater 26:221–238

16. Beija M, Alfonso CAM, Martinho JMG (2009) Synthesis and applications of Rhodamine derivatives as fluorescent probes. Chem Soc Rev 38:2410–2433
17. Avnir D, Levy D, Reisfeld R (1994) The nature of silica cage as reflected by spectral changes and enhanced photostability of trapped rhodamine 6G. J Phys Chem 88:5968–5958
18. Severin-Vantilt MME, Oomen EWJL (1993) The incorporation of Rhodamine B in silica sol-gel layers. J Non-Cryst Solids 159:38–48
19. Lee MH, Lee SJ, Jung JH, Lim H, Kim JS (2007) Luminophore-immobilized mesoporous silica for selective Hg^{2+} sensing. Tetrahedron 63:12087–12092
20. Jackson CL, Bauer BJ, Nakatani AI, Barnes JD (1996) Synthesis of hybrid organic–inorganic materials from interpenetrating polymer network chemistry. Chem Mater 8:727–733
21. Innocenzi P, Kidchob T, Yoko T (2005) Hybrid organic-inorganic sol-gel materials based on epoxy-amine systems. J Sol-Gel Sci Technol 35:225–235
22. Innocenzi P, Brusatin G, Guglielmi M, Bertani R (1999) New synthetic route to (3-Glycidoxypropyl)trimethoxysilane-based hybrid organic-inorganic materials. Chem Mater 11:1672–1680
23. Innocenzi P, Sassi A, Brusatin G, Guglielmi M, Favretto D, Bertani R, Venzo A, Babonneau F (2001) Chem Mater 13:3635–3643
24. Ayandele E, Sarkar B, Alexandridis P (2012) Polyhedral oligomeric silsesquioxane (POSS)-containing polymer nanocomposites. Nanomaterials 2:445–475
25. Cordes DB, Lickiss PB, Rataboul F (2010) Recent developments in the chemistry of cubic polyhedral oligosilsesquioxanes. Chem Rev 110:2081–2173
26. Shea KJ, Loy DA (2001) A mechanistic investigation of gelation. The sol–gel polymerization of precursors to bridged polysilsesquioxanes. Acc Chem Res 34:707–716
27. Cerveau G, Corriu RJP (1998) Some recent developments of polysilsesquioxane chemistry for material science. Coord Chem Rev 1051:178–180
28. Mehdi A (2010) Self-assembly of layered functionalized hybrid materials. A good opportunity for extractive chemistry. J Mater Chem 20:9281–9286
29. Livage J (1994) Sol-gel synthesis of transition metal oxopolymers. In: Frontiers of polymers and advanced materials. Springer, Boston, pp 659–667
30. Livage J, Henry M, Sanchez C (1988) Sol-gel chemistry of transition metal oxides. Prog Solid St Chem 18:250–341
31. Wen J, Wilke GL (1996) Organic/inorganic hybrid network materials by the sol-gel approach. Chem Mater 8:1667–1681

Chapter 3
From a Sol to a Gel

Abstract The transformation of an inorganic sol into a gel is a complex process which involves several reactions. Hydrolysis and condensation govern the process, but several parameters, such as the pH, the catalyst, the concentration of the precursor, and the temperature, are some of the variables that affect the formation of the final gel and the sol to gel transition.

Keywords Hydrolysis · Condensation · Esterification · Point of zero charge · Gelation time · Gravity

3.1 From a Sol to a Gel: Formation of a Reactive Monomer (Partial Hydrolysis)

The sol-gel process begins with the formation in the alkoxides of reactive groups [1]; they are necessary to proceed further with the condensation reactions which will form an oxide cluster. At first, the alkoxide is diluted in a solvent, generally the alcohol of the alkoxy group. This means, for instance, that if the alkoxide is tetraethyl orthosilicate (TEOS), the best choice is ethanol. It seems obvious, but if another alcohol is chosen for any specific reason, such as the need of a different boiling point, the alcohol exchange reactions which can affect the kinetics of the process have to be taken into account, as we have seen in the previous chapter.

In principle, the addition of water to the silicon alkoxide should generate Si-OH groups through hydrolysis reactions. A silicon alkoxide, however, reacts very slowly in water and alcohol, and the addition of a catalyst is necessary.

The need for this first step, which is the creation of reactive species through hydrolysis, is marking the difference with a typical organic polymerization. In the hydrolysis step, the hydroxyl groups (OH) replace, via a nucleophilic attack on the silicon atom by the oxygen atom of a water molecule, the alkoxide species (OR).

This process produces the release of an alcohol molecule and the formation of a metal hydroxide, M-OH (3.1):

$$M(OR)_n + H_2O \rightarrow (RO)_{n-1} MOH + ROH \qquad \textbf{Hydrolysis} \qquad (3.1)$$

where R is the alkyl group, ROH the alcohol, and M the metal. In the case of TEOS, the hydrolysis reaction becomes (3.2):

$$Si(OC_2H_5)_4 + H_2O \rightarrow (C_2H_5O)_3 SiOH + C_2H_5OH \qquad (3.2)$$

The reaction (3.1) can go, however, also in the other direction giving rise to *esterification*; alcohol can react with a hydrolyzed species forming a water molecule and an alkoxide ligand again (3.3):

$$M(OR)_n + H_2O \leftarrow (RO)_{n-1} MOH + ROH \qquad \textbf{Esterification} \qquad (3.3)$$

This means that alcohol has an active part in the process and does not play the simple role of solvent. On the other hand, because water and the alkoxysilanes are immiscible, we need a mutual solvent to homogenize the solution, even if in some cases the alcohol generated as hydrolysis by-product could be enough for this purpose.

Figure 3.1 shows the TEOS-ethanol-water ternary phase diagram [2]; in this system, the miscibility zone increases with the decrease of TEOS content as it should be expected because of the hydrophobic nature of the ethoxy groups in

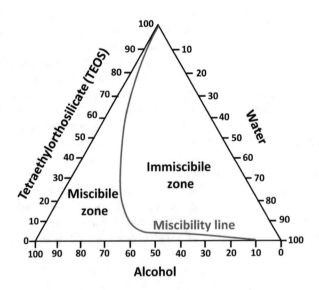

Fig. 3.1 Ternary phase diagram of the TEOS-ethanol-water system (25 °C). The alcohol has been added as a 95% ethanol and 5% water mixture

TEOS. The choice of the co-solvent is not limited to ethanol, and other alcohols or solvents, such as tetrahydrofuran or 1,4-dioxane, have been employed for sol-gel syntheses.

Another point to stress is that we should not think to this first step as a very well-defined stage; in fact, as soon as hydrolysis also starts, condensation begins.

The necessity of adding a catalyst produces a first effect, which is the change of pH of the solution as a function of the type and amount of the catalyst. We can have, therefore, basic or acidic catalytic conditions which affect the condensation reactions of silica very much. To make this point more clearly, we can use the concept of the *point of zero charge*.

3.2   The Point of Zero Charge

The *point of zero charge* is used to indicate the condition when the electrical charge density on a surface is zero. If ionic groups cover the surface of a particle, the counterions in the solution will cover this layer balancing the charge. The pH value of the particle becomes neutral in the *point of zero charge* (PZC). The surface charge depends, therefore, on the pH:

$$pH > PZC \rightarrow \text{the surface is } \textbf{negatively } \text{charged}$$

$$pH < PZC \rightarrow \text{the surface is } \textbf{positively } \text{charged}$$

The PZC in oxides depends on the preparation and measurement method; some reference values are [3] TiO_2 (anatase) 4.2, SnO_2 3.5, ZrO_2 5.5, Stöber silica 2.1, and γ-Al_2O_3 8.5. The PZC of silica containing silanol groups typically ranges between pH 1.8 and 4.2. This value during a sol-gel reaction will change accordingly with the degree of silica condensation.

In the case of hydroxides, the surface potential is given by the balancing of the H^+ and OH^- ions; the charge of the surface is pH dependent:

$$M-OH + H^+ \rightarrow M-OH_2^+ \quad (pH < PZC) \tag{3.4}$$

$$M-OH + OH \rightarrow M-O + H_2O \quad (pH > PZC) \tag{3.5}$$

In the two extremes of pH, below 2 and higher than 13, the hydrolysis will be very fast. In these conditions, however, the sol will tend to stabilize because the particles with the same charge will repel each other, the sol is quite stable, and condensation is hindered or very slow.

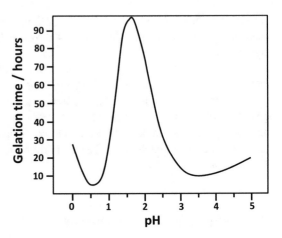

Fig. 3.2 Gelation time as
a function of pH for a
silica (TEOS) acid-
catalyzed (HCl) sol

3.2.1 The PZC and the Gelation Time

The pH and the PZC also have an important effect on the reaction rates and, therefore, the gelation time; Fig. 3.2 shows how the gel time of an acid-catalyzed silica
sol (TEOS, HCl with H_2O/TEOS = 4) is affected by the change of pH in acid conditions. The gelation time as a function of pH shows a response which looks like a
Gaussian curve. The longest gel time is observed around pH = 2.2, which also corresponds to the PZT of the system. At higher values a fast decrease in the gel time
occurs as such as at low pH; at very low pH, it is observed a small increase which is
due to the dissolution of silica in high acid conditions.

The PZC of silica corresponds, therefore, to the highest temporary stability of
the sol.

3.3 From the Precursor to a Sol, the Transition Metals

In the case of titanium, but it can be generally extended to other transition metals,
the nucleophilic addition of water to the Ti center is the mechanism at the base of
the hydrolysis. When titanium is dissolved in water as a salt, the T^{4+} cations (or any
cation M^{z+}, of charge z) are solvated by the water molecules (3.6):

$$\text{Ti}^{4+} + :\overset{\displaystyle /H}{\underset{\displaystyle \backslash H}{O}} \rightarrow \left[\text{Ti}:\overset{\displaystyle /H}{\underset{\displaystyle \backslash H}{O}} \right]^{4+} \qquad (3.6)$$

This produces a charge transfer from the oxygen to the metal atom with a con-
temporary increase of the partial charge of hydrogen; the water molecules coordinated with the metal ions are more acidic than those not coordinated. The extent of

hydrolysis is, therefore, depending on water acidity and the entity of the charge transfer up to reaching the equilibria (3.9):

$$\left[Ti\left(OH_2\right)\right]^{4+} \rightleftharpoons \left[Ti\left(OH\right)\right]_3 + H^+ \rightleftharpoons \left[Ti = O\right]^{2+} + 2H^+ \tag{3.9}$$

In non-complexing aqueous media, therefore, following Eq. 3.9, three types of different ligands would form [4, 5]:

$$Ti \ \left(OH_2\right)\left(Aquo\right) \quad Ti \ OH\left(Hydroxo\right) \quad Ti = O\left(Oxo\right)$$

Metal cation charge and pH are the two parameters that regulate the extent of the three domains, *aquo*, *hydroxo*, and *oxo*, as shown in the "charge-pH" diagram [6] in Fig. 3.3:

The diagram gives direct visual information about what kind of species would form as a function of the pH and the charge of the metal cation, M^{z+}. In general for atoms with z charge lower than 4, only aquo or hydro species can form for the whole range of pHs. At higher charges, $z > 5$, *hydro* or *oxo* species are observed, while at the intermediate value $z = 4$, all the possible species can form by a proper change of pH.

Because the condensation goes through the reaction of hydroxo groups with water elimination (3.10):

$$\equiv Ti - OH + OH - Ti \equiv \ \rightarrow \ \equiv Ti - O - Ti \equiv + \ H_2O \tag{3.10}$$

Equation 3.9 has to be shifted to the Ti-OH region by controlling the pH.

In the case of titanium alkoxides [7], and more in general transition metal alkoxides, because they are stronger Lewis acids than silicon, a nucleophilic attack is easier, and this results in a higher hydrolysis rate. The condensation goes through the M-OH reaction which can be so fast that almost immediate precipitation can be observed upon addition of water [8].

Fig. 3.3 The *aquo*, *hydro*, and *oxo* domains as a function of the charge and pH

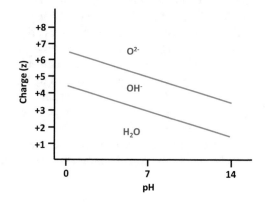

3.4   From a Sol to a Gel: Condensation

As soon as the first reactive –OH groups are produced, the condensation reactions will also start to take place. In silica systems, the condensation forms -Si-O-Si- units by releasing water or an alcohol molecule. The reaction of two -Si-OH groups will give water as a by-product (Eq. 3.11), while the reaction of -Si-OH with -Si-OR will release an alcohol molecule (Eq. 3.12):

$$-Si-OH + -Si-OH \rightarrow -Si-O-Si- + H_2O \qquad (3.11)$$

$$-Si-OH + -Si-OR \quad \rightarrow \quad -Si-O-Si- + ROH \qquad (3.12)$$

With the progress of the polycondensation reactions, an extended oxide network forms whose structure and growth depend on a set of synthesis parameters. We have seen that water is not enough to start the hydrolysis and condensation reaction, and we need a catalyst. The choice of the catalyst is very important because the sol structure depends mainly on this selection, and therefore also the gel time and the sol to gel transition will be affected [9]. The first choice is between an acidic and a basic route for sol-gel processing; so much of the final structure of the gel would depend on this parameter.

3.5   Acid-Catalyzed Hydrolysis and Condensation

To understand why the choice of the catalyst is so important, it is better to have a look a little more in depth to the mechanism of the involved reactions, without in any case going too much in detail. In acid catalysis, the protons which are available in solutions will try to find electrons, and the best place to get them is the oxygen atom of the Si-OR group. After the attack of the proton, the electronic cloud in the Si-O bond will be shifted from silicon to oxygen, and this will increase in turn the positive charge of the silicon atom. At this point, silicon becomes more electrophilic and more reactive to the attack of water during the hydrolysis; the same will happen to silanols in the condensation reactions. The higher electrophilicity of silicon induced by the protonation also has another effect, which is a change of its reactivity. The unreacted alkoxide $(Si-(OR)_4)$ hydrolyzes faster than the partially hydrolyzed $(Si(OR)_{4-x}(OH)_x)$ or condensed silica (-Si-O-Si-). This is an interesting point to catch because the more the hydrolysis and condensation reactions continue, the more the pH of the sol will change. The silanol groups become more acidic with the increase of condensation when more Si-O-Si bonds are present; this is also reflected in a change of PZC with the progress of the condensation.

Acid Catalysis

Protonation of ≡Si-OR

$$\equiv Si\text{-}OR + H^+ \rightleftharpoons \equiv Si\text{-}O\overset{\cdots H}{\underset{R}{}} \longrightarrow \equiv Si\text{-}O\overset{+\cdots H}{\underset{R}{}} + HOH \longrightarrow \equiv Si\text{-}OH + ROH$$

Hydrolysis

Protonation of ≡Si-OH

$$\equiv Si\text{-}OH + H^+ \rightleftharpoons \equiv Si\text{-}O\overset{\cdots H}{\underset{H}{}} \longrightarrow \equiv Si\text{-}O\overset{+\cdots H}{\underset{R}{}} + Si\text{-}OH \longrightarrow \equiv Si\text{-}O\text{-}Si \equiv$$

Condensation

3.6   Basic Catalyzed Hydrolysis and Condensation

In the case of basic catalyzed hydrolysis and condensation, the reactions are caused by hydroxyl ions (OH⁻) which have strong nucleophilicity and are strong enough to attack the silicon atom directly. Silicon in the alkoxide is the atom which is carrying the highest positive charge and becomes, therefore, the target of the nucleophilic attack from deprotonated hydroxyls (OH⁻) or silanols (≡Si-O⁻). In the basic catalyzed reaction, OH⁻ and ≡Si-O⁻ species will replace OR (hydrolysis) or ≡Si-OH (condensation), respectively. The associative mechanism involves the formation of a pentacoordinate intermediate.

Basic Catalysis

Deprotonation of water

$$\equiv Si\text{-}OR + HO^- \rightleftharpoons \left[\underset{\equiv Si\text{-}OR}{OH} \right]^- \rightleftharpoons \equiv Si\text{-}OH + RO^-$$

Hydrolysis

Deprotonation of silanols

$$\equiv Si\text{-}OH + \equiv SiO^- \rightleftharpoons \left[\underset{\equiv Si\text{-}OH}{SiO^-} \right]^- \rightleftharpoons \equiv Si\text{-}O\text{-}Si \equiv + HO^-$$

Condensation

The condensation reaction represented just above of course can happen also for any alkyl group R at the place of H. These reactions in high basic conditions are also reversible via cleavage by OH⁻.

3.7   How Much Water?

We have seen that water is an essential component in the sol-gel synthesis and one of the key parameters to be controlled in the process. Not hydrolytic routes are also feasible, and specific sol-gel chemistry has been developed with good results.

The different stages of the sol to gel transition are strongly dependent on the water/alkoxide ratio, r (alternatively is also used the alkoxy/water ratio, R_w); clearly, the amount of water which is available to start the hydrolysis will affect the kinetics of the polycondensation process. The stoichiometric value of r is 4 ($R_w = 1$); it means that four molecules of water are necessary for complete hydrolysis of a tetravalent alkoxide $M(OR)_4$, while a ratio of 2 is enough for conversion of $M(OR)_4$ into an oxide.

What should be expected is that if we increase the amount of water available for hydrolysis, also the polycondensation rate should increase. It is not really true because if we add more water while keeping constant the amount of solvent, the silicate concentration decreases. This dilution effect will change the hydrolysis and condensation rate with an increase in the gel time.

Figure 3.4 shows how the gel time changes as a function of the water/TEOS ratio keeping constant the ethanol which is used as a solvent. The gray and the white areas in the figure mark the different response observed when water is present in under stoichiometric amount, $r < 4$ (gray area), or higher, $r > 4$ (white area). As soon as we increase the content of water, the gel time decreases because more water is available for the hydrolysis; after around $r = 5$, however, the dilution effect is more effective, and the gelation time increases quite quickly with the water content. The gel time also increases with the amount of ethanol in the sol. The concentration of the oxide species is very important; the more the sol is diluted, the longer will be the gel time. Another question which arises with the increase of water is that the system could potentially enter in an area of immiscibility in the ternary phase diagram

Fig. 3.4 The gel time as a function of the water/ TEOS molar ratio. The three different curves show the change of gel time at different ethanol/TEOS ratios (1, 2, and 3) [10]

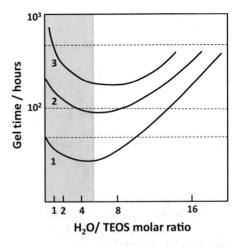

water/TEOS/ethanol. The polycondensation reactions, however, produce also alcohol as a by-product which in most of the cases is enough to homogenize the system.

3.8   Hydrolysis vs Condensation

During hydrolysis and condensation reactions, the silicon alkoxides undergo a transformation through a transition state. The electronic density of the silicon atom, as we have seen, also depends on the nature of the substituents and decreases with the progress of the reactions in the following order (3.13):

$$\equiv Si - OR > \ \equiv Si - OH > \ \equiv Si - O - Si \equiv \tag{3.13}$$

The decrease in electron density of silicon during acid-catalyzed reactions has the consequence that in acidic conditions also the reaction rates of hydrolysis and condensation increase in the order of the electronic density; the higher is the electronic density (Si-OR), the higher is also the hydrolysis rate.

This means that in acid-catalyzed systems, the hydrolysis is faster than condensation; in basic catalyzed sols instead a reverse trend is observed [11]. Besides the differences in the reaction rate, the basic and acidic routes also produce a more subtle difference which is the structure of the silica clusters. In acid conditions, because of the higher reactivity of the electrophilic silicon atom, with the growth of -Si-O-Si- bonds, more chain-like structures are favored. On the turn, in basic conditions branched and more connected silica species are obtained.

The organically modified silicon alkoxides, R'(SiOR)$_3$, have a higher electron density at the silicon atom, and Eq. 3.13 can be rewritten (3.14):

$$\equiv Si - R' > \ \equiv Si - OR > \ \equiv Si - OH > \equiv Si - O - Si \equiv \tag{3.14}$$

This means that, in comparison with a silicon alkoxide, they have in acidic conditions a higher reactivity which increases with the number of organic substituents and in the case of R' = CH$_3$ (3.15):

$$\left(CH_3\right)_3 - Si - OCH_3 > \left(CH_3\right)_2 - Si - \left(OCH_3\right)_2 > \left(CH_3\right) - Si - \left(OCH_3\right)_3 \tag{3.15}$$

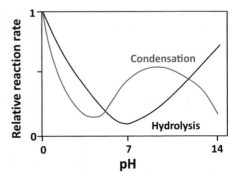

Fig. 3.5 Relative hydrolysis and condensation rates as a function of pH for a silicon alkoxide

The reactivity of organically modified alkoxides also depends on the steric hindrance of the organic substituent groups and increases in the order (3.16):

$$MTES > VTES > TEOS \tag{3.16}$$

In basic conditions, the reaction rates of hydrolyzed or partially hydrolyzed species are higher than the monomeric alkoxide, the opposite for what we have seen in the case of acid catalysis. This is also true for the organically modified alkoxides which in basic conditions react slower than the corresponding silicon alkoxide.

The pH of the solution will trigger, therefore, the reaction rates of hydrolysis and condensation which remain competing reactions for all the sol-gel process.

The change of hydrolysis and condensation rates as a function of pH can be followed in Fig. 3.5. At pH lower than around 5, the hydrolysis rate is faster than condensation, as it could be expected; the hydrolysis rate also decreases with the increase of pH and reaches a minimum around 7. After this value, the hydrolysis rate increases quite quickly with the alkalinity of the solution. The condensation rate, on the other hand, follows a similar trend and decreases with the increases of pH even if it has a lower reaction rate with respect to hydrolysis up to the value of around 5. After this pH value, the condensation rate quickly rises to 10 and then decreases again. Why is observed this trend? Because we should always keep in mind that cleavage of silica bonds at higher pH values is quickly rising, and condensation and hydrolysis are in competition.

Condensation and hydrolysis reactions are also depending on the size of the alkoxy group [12]; the reactivity of silicon alkoxides, in fact, decreases with the increase of the size of the alkoxy because of the steric hindrance. The reactivity of silicon alkoxides decreases when the size of the alkoxy group increases because of steric hindrance factors. The reaction rate order of silicon alkoxides with different alkoxy groups follows this order:

$$Si\left(OMe\right)_4 > Si\left(OEt\right)_4 > Si\left(O^nPR\right)_4 > Si\left(O^iPR\right)_4 > Si\left(O^nBu\right)_4 > Si\left(OHex\right)_4 \tag{3.17}$$

So now we could ask which is the better choice, acid or basic synthesis? Well, it depends of course on the type of final material we have in mind. The different

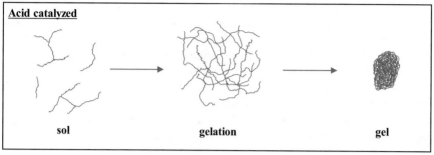

Acid catalyzed hydrolysis

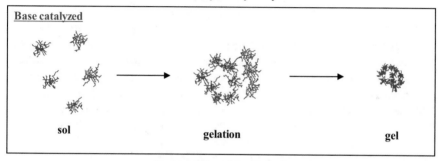

Base catalyzed hydrolysis

Fig. 3.6 Formation of a gel from an acid (**a**) or basic (**b**) catalyzed hydrolysis, the change of inorganic structure from a sol to a gel

hydrolysis-condensation rates in acid and basic conditions have a direct effect on the structure of the gel and therefore of the dried material. We have just seen that in acid conditions, linear or weakly branched silica species are preferentially formed; they aggregate through entanglements which eventually leads to gelation. In acid conditions gelation occurs via formation of agglomerated silica clusters which condense to form a 3D network (Fig. 3.6). The acid conditions give rise, therefore, to a final material which is denser concerning the basic route which forms a material with a more porous network because of the free space between the particles. A dense silica material is obtained only after firing at high temperatures, but in the gel or xerogel state, the structural differences between acid and silica gels are still important.

3.9   Formation of the Sol

The hydrolysis and condensation reactions start as soon the catalyst is added to the solution; from this moment an interconnected silica network begins to form. The aggregates soon will reach a colloidal dimension (sub-micrometer), and finally, a sol is obtained. The structure of these particles and their density depend on the synthesis parameters, such as the pH and the water alkoxide ratio, as we have just seen. The structure of these particles also depends on the type of catalysis; base-catalyzed systems are characterized by ramified structures, while acid-catalyzed systems will form more linear structures with small branching.

How do the particles of the silica sol form? A simple answer to this question is difficult to give because a main characteristic of the sol-gel chemistry is the complexity of the possible reactions pathways from the very beginning of the process. Several types of intermediate species form from the very beginning of the sol-gel process, and each one of these intermediates can hydrolyze and/or condense. Linear, branched, cyclic, and cage silica species can form and dissolve throughout the process.

From the catalysis of the monomeric alkoxide precursor, reactive monomers (the hydrolyzed alkoxide), dimers, trimers, and linear and cyclic species (silica rings and cages) can be obtained in the order (see Fig. 3.7). The rings become a kind of nucleation center, and via the addition of monomers and other species, three-dimensional particles can finally form. Different types of silica rings can form, from twofold to sixfold rings; in general however only the structures with at least four silicon tetrahedra tend to be stable. Cyclization can be intramolecular, by forming a closed loop by condensation of reactive sites within the same molecule or more likely intermolecular (see Chap. 5). Chain silica structures are more stable than cyclic one and the most likely to form at the very early stages of hydrolysis and condensation but as soon as structures with three or four tetrahedra form the cyclic structures are favored and particles of 1–2 nm are formed [13].

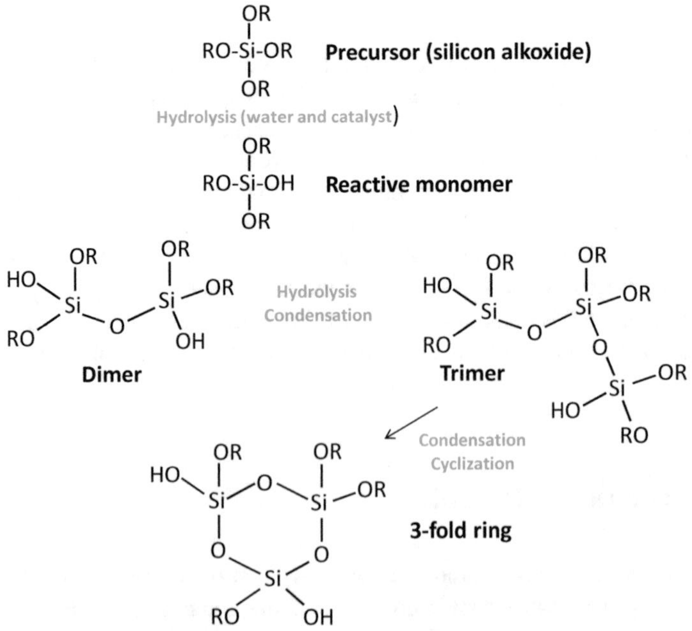

Fig. 3.7 Formation of cyclic species during sol-gel reactions

After the formation of the first silica, clusters form, the particles grow by Ostwald ripening mechanism, which means that the largest particles, which are less soluble, grow at the expense of the smallest one.

We will see in the next chapter that in theoretical models, such as the classic theory, the formation of cyclic species is neglected. However, they play a critical role during the sol to gel transition [14].

An example is what happens during gelation of bridged silsesquioxanes, which show a discontinuity in the process. Monomers which contain one to four bridging carbon atoms show a high tendency to fast cyclization. The cyclic intermediates become "local thermodynamic sinks" which hamper the condensation and growth process with formation of "kinetic bottlenecks." When cyclic species form, gelation time increases, and in some cases the sol to gel transition is completely shut down. Cyclization, in fact, "reduces the level of functionality and does not contribute to network formation."

Figure 3.8 shows how the gelation time changes with the increase of the chain length of alkylene-bridged monomers under acidic conditions. Three different types of responses as a function of the length of the organic bridging chain are observed. If the chain is composed of only one or two methylenes, the gelation time is slower. It increases very much instead when three or four bridging groups are present. When the number of bridging carbon atoms is higher than four, the gelation time is very fast. Cyclization of shorter chain bridged monomers is responsible for the deviation of the gelation time.

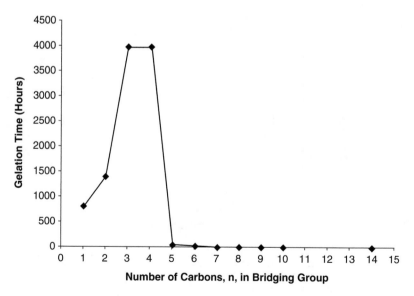

Fig. 3.8 Graph of gelation times for alkylene-bridged polysilsesquioxanes under standard sol-gel conditions with HCl. Under acidic conditions, small changes in the bridging alkylene chain length produce profound changes in gel time. (Reproduced with permission from ref. Cordes DB, Lickiss PB, Rataboul F (2010) Recent developments in the chemistry of cubic polyhedral oligosilsesquioxanes. Chem. Rev 110: 2081–2173.)

3.10   The Effect of Container Size on Gelation Time

We have seen that the gelation time is directly dependent on a direct number of parameters, such as temperature, pH, water amount and dilution, etc. But are we sure that we have considered everything? Maybe no, because there is another parameter which is generally overlooked and that is instead very important, the size of the sol container [15]. It can be quite surprising, but both the theoretical calculations and the experimental observations demonstrate a box-size dependence of the gelation time, t_g (Fig. 3.9). In the case of a basic catalyzed TMOS sol, the t_g has been observed to increase with the size of the container and the decrease of the volume fraction. A similar effect has also been observed by other researchers in acid-catalyzed sols [16]. This means that the more the precursor sol is diluted, the more the container size effect has to be taken into account. The theoretical simulations are in good agreement with these experimental results.

If the container size is important, what about the material of the container? If the box is done in glass or polymeric materials, it is reasonable also to expect some changes of the chemical reactivity of the sol at the container interface. In conclusion, scale-up of sol-gel reactions is quite tricky, and all the experimental conditions, including the size and the material of the container, have to be considered. Another recommendation is that the experimental section of an article should include an accurate description of the vessels, size and material, used for sol-gel reactions.

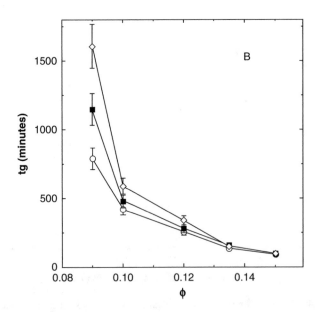

Fig. 3.9 The experimental dependence of the gel time, t_g, on container size and sol volume fraction, ϕ, of TMOS. Hollow circles 10 mm, squares 14 mm, and hollow diamonds 32 mm

3.11 What About Gravity?

The experimental data used to analyze the sol to gel transition have been all col-
lected at the terrestrial gravity. What could happen to realize the sol-gel process at
different gravity conditions? The question is quite interesting because we could
reasonably expect that the gravity affects the reactivity of the species and therefore
the final structures. Is this true? The answer has been given by some experiments
realized at different gravity conditions to study how the final gel structure is affected,
and some interesting differences as a function of the gravity level have been
observed. Changing the mobility (diffusion) of the species in the sol would also
change the gel structure [17].

In microgravity conditions, the diffusion is limited (*diffusion-limited growth*),
and this favors intramolecular rather than intermolecular reactions [18].
Intramolecular cyclization would become the dominant process in silica sols. At
gelation, the silica structure is formed by interconnected cyclic building units
(Fig. 3.10). A different situation is observed in increasing gravity. Intermolecular
reactions compete with intramolecular cyclization, and a mixture of branched and

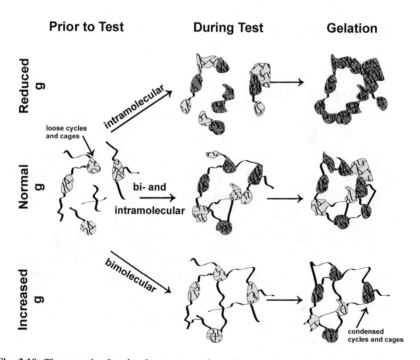

Fig. 3.10 The growth of molecular structures in various gravity regimes. In reduced gravity, a
diffusion-limited regime results in domination of intramolecular cyclization. The cyclic building
units then react with their nearest neighbors, leading to extended coils of cages and a structure high
in Q4 groups at gelation. In normal and high gravity, diffusion allows bimolecularization to compete with cyclization, resulting in a mixture of chains as well as cycles. (Reprinted with permission
from Ref. [17])

cyclic structures forms the final structures of the silica gel. Gravity has also been found to affect the formation of silica particles because of the diffusion-limited aggregation [16].

The gravity does not affect only silica sols which we know have a very peculiar reactivity with respect to other oxides. By studying TiO_2 sol-gel reactions in reduced, terrestrial, and high gravity, it has been observed that because of the removal of downward convection, hydrolysis reactions are favored with respect to condensation. Noticeably, making the reactions at high gravity conditions, crystalline anatase materials could be obtained afterthermal treatment at 40–50 °C, instead, the much higher temperatures (above 400 °C) required by conventional methods [19].

3.12   A General View

We can finally summarize the overall process, from sol to gel, using the simple but effective "classic" scheme of structural development proposed by Iler [20] in his book on silica chemistry (Fig. 3.11).

A three-dimensional gel network is obtained via hydrolysis of the precursor (monomer) which condenses to form dimers and trimers and cyclic species.

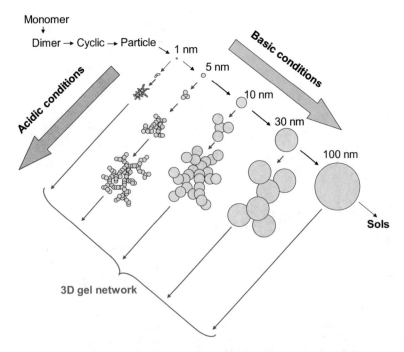

Fig. 3.11 Scheme of silica gel formation. (Redrawn with permission from Ref. [19]

Condensation via acid catalysis of the first oligomers gives branching and linear species, while the basic route mainly produces small nanoparticles which aggregate to form a gel or eventually grow in size.

References

1. Osterholtz FD, Pohl ER (1992) Kinetics of the hydrolysis and condensation of organofunctional alkoxysilanes: a review. J Adhes Sci Technol 6:127–149
2. Coyan HD, Setterstrom CA (1946) Properties of ethyl silicate. Chem Eng News 24:2499
3. Kosmulski M (2009) pH-dependent surface charging and points of zero charge. IV. Update and new approach. J Colloid Interface Sci 337:439–448
4. Rozes L, Sanchez C (2011) Titanium oxo-clusters: precursors for a Lego-like construction of nanostructured hybrid materials. Chem Soc Rev 40:1006–1030
5. Rozes L, Steunou N, Fornasieri G, Sanchez C (2006) Titanium-oxo clusters, versatile nanobuilding blocks for the design of advanced hybrid materials. Monatshefte fur Chemie 137:501–528
6. Livage J, Henry M, Sanchez C (1988) Sol-gel chemistry of transition metal oxides. Prog Solid State Chem 18:250–341
7. Yoldas BE (1986) Hydrolysis of titanium alkoxide and effects of hydrolytic polycondensation parameters. J Mater Sci 21:1087–1092
8. Cargnello M, Gordon TR, Murray CB (2014) Solution-phase synthesis of titanium dioxide nanoparticles and nanocrystals. Chem Rev 114:9319–9345
9. Pope EJA, Mackenzie JD (1986) Sol-gel processing of silica. II. The role of the catalyst. J Non-Cryst Solids 87:185–198
10. Klein LC (1985) Sol-gel processing of silicates. Annu Rev Mater Sci 15:227–248
11. Brinker CJ (1988) Hydrolysis and condensation of silicates: effects on structure. J Non-Cryst Solids 100:31–50
12. Hook R (1996) A ^{29}Si NMR study of the sol-gel polymerisation rates of substituted ethoxysilanes. J Non-Cryst Solids 195:1–15
13. Ng LV, Thompson P, Sanchez J, Macosko C, McCormick AV (1995) Formation of cagelike intermediates from nonrandom cyclization during acid-catalyzed sol-gel polymerization of tetraethyl orthosilicate. Macromolecules 2:6471–6476
14. Depla A, Verheyen E, Veyfeyken A, Van Houteghem M, Houthoofd K, Van Speybroeck V, Waroquier M, Kirschhock CEA, Martens JA (2011) UV-Raman and ^{29}Si NMR spectroscopy investigation of the nature of silicate oligomers formed by acid catalyzed hydrolysis and polycondensation of tetramethylorthosilicate. J Phys Chem C 115:11077–11088
15. Anglaret E, Hasmy A, Jukkien R (1995) Effect of container size on gelation time: experiments and simulations. Phys Rev Lett 65:4059–4062
16. Huber CJ, Butler RL, Massari AM (2017) Evolution of ultrafast vibrational dynamics during sol-gel aging. J Phys Chem C 121:2933–2939
17. Smith DD, Sibille L, Cronise RJ, Hunt AJ, Oldenburg SJ, Wolfe D, Halas NJ (2000) Effect of microgravity on the growth of silica nanostructures. Langmuir 16:10055–10060
18. Pienaar CL, Chiffoleau JA, Follens LRA, Martens JA, Kisrchhock EA, Steinberg TA (2007) Effect of gravity on gelation of silica sols. Chem Mater 19:660–664
19. Hales MC, Steinberg TA, Martens WN (2014) Synthesis and characterization of titanium sol-gels in varied gravity. J Non-Cryst Solids 396–397:13–19
20. Iler RK (1979) The chemistry of silica. John Wiley & Sons, Inc., New York

Chapter 4
Sol to Gel Transition: The Models

Abstract The sol to gel transition is a critical phenomenon which is observed in systems of different nature and composition. Several theories using different mathematical approaches have been developed with the main purpose of predicting the gel point. These models have been applied to systems which grow in a stochastic way and to different practical problems. Most of these theories have been used for studying gelation in organic polymers and colloidal systems of particles, but they also have been, even if a much smaller extent, applied to inorganic sol-gel systems. Some of these models reach a quite good correspondence with experimental results; in other cases even if they fail to match with the experimental data, they are however an important tool for a basic understanding of the process. The models are also used to make previsions and to find answers to some fundamental questions: when the system will gel, which is the minimum number of reactions (bonds formation) to observe a sol to gel transition?

Keywords Flory-Stockmayer theory · Percolation · Bond percolation · Site percolation, Scaling laws

4.1 The Classic Statistical Model (Mean-Field Theory)

One of the most important theories for describing polymerizing systems is the *classic statistical theory* (also known as *mean-field theory*) which has been developed by Flory and Stockmayer (FS) [1, 2]. The model considers how a monomer with a specific functionality can form an interconnected structure (a gel) without forming rings. The model is based on some assumptions; not all of these assumptions fit well with the real case, and this represents the limit of the classical model. The theory, however, has the merit to give a first general answer to one of the previous questions, how many bonds we need to form a gel?

© The Author(s), under exclusive license to Springer Nature Switzerland AG 2019
P. Innocenzi, *The Sol-to-Gel Transition*, SpringerBriefs in Materials,
https://doi.org/10.1007/978-3-030-20030-5_4

The assumptions at the basis of the model are:

− The reactivity of all the functional groups of a monomer remains the same during the overall polymerization process. (From the previous chapters, we know, however, that is not the case in a sol-gel process; in fact with the proceeding of hydrolysis and condensation, the reactivity of the different species (Si-OR, Si-OH and Si-O-Si) changes.)
− Connecting bonds form only between different monomers or polymers, and no ring formation is therefore allowed. (Formation of cyclic species plays instead an important role in the sol-gel process.)
− Volume and steric hindrance effects are considered negligible.

Using these hypotheses the polymerization in the classical model can be visualized by a graph called the Cayley tree; the name is due to the similarity with a forest tree whose branches do not have cyclic interconnections. Another graph commonly used is the Bethe lattice, which corresponds to the "infinite regular Cayley tree" [3]. The graph is drawn using the previous assumptions, the functionality of the monomer, or a maximum number of allowed bonds, f, the number of nodes, n, and the total number of bonds, b; the bond formation probability, p, depends on these parameters (4.1):

$$p = \frac{b}{fn} \tag{4.1}$$

In the classic theory, the probability p that a monomer can form one of the possible f bonds does not depend on how many of the other $f - 1$ bonds have been already formed.

If we consider the case of a silicon alkoxide, the functionality will be $f = 4$.

Figure 4.1 shows a Cayley tree model in the case of a monomer with $f = 4$, such as a silicon alkoxide. The total number of nodes in Fig. 4.1 is 21, while the bonds are 40. To get the total number of bonds, we have to consider all the connections for every node, which means that some of them need to be counted two times. For instance, the node 3 has 4 bonds, the node 12 has 1 bond, while node 5 has 2 bonds.

With the help of Eq. 4.1, we can now calculate the probability, p, to have a connection at each site:

$$p = \frac{40}{4 \cdot 21} = 0.49 \tag{4.2}$$

Fig. 4.1 The Cayley tree for $f = 4$, $n = 21$, and $b = 40$. The probability p is 0.49

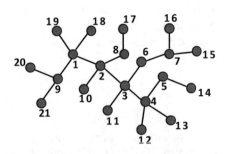

All the nodes in the examples are interconnected from one side to the other which means that a macromolecule of this type can form a gel. The other observation is that every node should have at least two connections to form an interconnected molecule (gel). This is generally true, but if we look carefully, it is clear that some molecules have only one bond, all the monomers at the boundary. This is a boundary effect, and in general, in a three-dimensional larger cluster, only the central part is considered representative of the whole system, neglecting surface effects.

At this point it is possible to define the *critical probability*, p_c, to observe the formation of a gel. The probability for a molecule of functionality, f, of forming another bond with a neighboring molecule is p. This is what in the Flory theory is defined as the branching coefficient [4–6]. After the formation of a bond, the functionality becomes $(f - 1)$, and the probability to form another bond is now $p(f - 1)$. This process can be iterated and extended to all the directions of the growing network. In average if the number of bonds is less than 1, "finite-sized" branched structures would form, while on the opposite, if $p(f - 1) > 1$, an "infinite-sized" branched network would grow. The intermediate value, $p(f - 1) = 1$, which represents a transition point between the two situations, is exactly the critical probability, p_c, at the gel point.

$$p(f-1) < 1 \rightarrow \text{finite sized branches structures}$$

$$p(f-1) > 1 \rightarrow \text{infinite sized branched network}$$

$$p_c = \frac{1}{f-1} \rightarrow \text{critical probability, "gel point"}$$

In the case of a molecule with functionality $f = 4$, the value of p_c becomes 0.33 (4.3):

$$p_c = \frac{1}{4-1} = \frac{1}{3} \tag{4.3}$$

If $f = 3$, then $p_c = \frac{1}{2} = 0.5$, while for $p_c \geq 1$, no gelation in the system will be observed [7]. This tells us that, in the case of a molecule with functionality 4, by the simple assumptions behind the classic theory, *a gel forms when at least one-third of the reactions have occurred*. If the molecule has a functionality 3, instead, a higher relative number of bonds need to be formed, and the reaction of half of all the possible bonds are, in fact, necessary to have a gel. It should be underlined again that this applies to the case of polymerization of monomers all having the same functionality and the same type of functional groups.

In the case $f = 4$, at the gel point, two-thirds of the connecting bonds are still available for further reactions; we have therefore found an answer to the question "which is the minimum number of reactions (bonds formation) to observe a sol to gel transition." The minimum bond conversion to observe a sol to gel transition in an inorganic system is around 33% by the classic FS theory. This result is not corresponding to the experimental findings, which are instead even quite far from this value. The model gives a qualitative evaluation of the entity of the reactions involved in an ideal sol to gel transition. It is a good starting point for a critical understanding of the relationship between sol structure and gelation.

The *classic model* is also able to provide a qualitative prediction of the average molecular weight distribution of a sol which is going through the gelation process. At the beginning of the process, the most common species are the monomers which form, with the progress of the reactions, molecules of larger dimensions and higher average molecular weight. The distribution of these molecules changes as much we are getting closer to the gel point because of the fraction of molecules with the higher molecular weight increases at the expenses of monomers and smaller aggregates. At the gel point, however, as soon as the largest aggregates bond to the spanning cluster is reasonable to expect a drop in the fraction of molecules with higher molecular weights in the residual sol. Figure 4.2 shows the change of weight fraction (w_x) of aggregates formed by x monomeric units in a sol of tetrafunctional molecules, such as TEOS, as a function of condensation degree, p. The weight fraction of the gel is w_g, which for $p > p_c$ is $w_g = 1 - w_s$, with w_s the weight fraction of the sol. Before gelation, the monomers are the predominant species which quickly decreases close to the gel point. However, as we expected, after gelation (at p_c) the monomers in the sol are still the predominant molecular fraction as the molecules with the highest molecular weight join the gel-solid cluster.

The classical theory, even with the limitations given by neglecting the cyclic species, can provide previsions about two basic features of gelation: the weight fractions of the different reacting species and the critical condition (degree of reaction) for gel formation [8].

If we admit, however, that cyclic species cannot form in the gel, which is not the real case, the classic theory fails to give a physically congruent model (see Chap. 5). The prohibition of closed loops generates a new effect: a crowding problem.

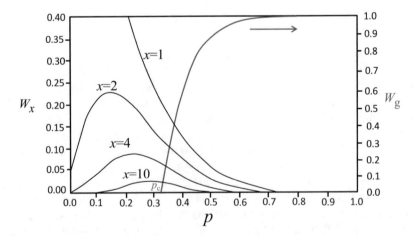

Fig. 4.2 Weight fraction (w_x) of aggregates formed by x monomeric units in a sol of tetrafunctional molecules as a function of reaction degree, p. w_g is the weight fraction of the gel

Fig. 4.3 A Bethe lattice

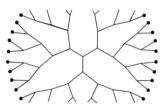

As soon as the polymer structure grows, the density also increases without limits. Figure 4.3 shows an example of Bethe lattice, with crowding at the boundaries due to the repeated branching. In such a structure, the mass, M, at the boundaries increases with the fourth power of the radius, r ($M \propto r^4$). Because the volume, V, increases with r^3, a comparison of the two proportions should give the conclusion that the density, ρ, increases with r ($\rho \propto r$), which is not a realistic description of a polymeric growth.

4.2 The Percolation Theory

The phase transition from a sol to a gel can also be treated using a mathematical approach; this is the case of *the percolation theory*, a model which has been applied to many different systems. This theory, even if it has not an analytical solution, is a very elegant model to describe a phase transition. What does it mean no analytical solutions? It means that it will be possible to provide only a statistical or probabilistic solution which is based on numerical calculations; the more powerful will be our computer and accurate the model, the more the numerical simulation will be able to give an accurate prevision.

The application of percolation theory to the sol to gel transition, more than for its practical application, is a very nice way to visualize and understand the random growth process which is at the ground of the connectivity transition and making it more intuitive.

The main difference with the classic theory, which we have just described, is that rings or closed loops are allowed to form.

4.2.1 Percolation: What Is it?

The slow motion and filtering of a fluid through a porous medium are generally defined as *percolation*. Coffee filtering is a common example to get an easy visualization of percolation; hot water percolates through a powdered coffee layer to extract the soluble component. High pressure-induced percolation is what is behind a good espresso. The water molecules have in any case to find their path to the bottom of the coffee layer, which acts as a porous filtering membrane. This is why the

Caffè moka   Caffè espresso   Caffè filtro   Caffè alla napoletana

Fig. 4.4 Coffee can be prepared by different percolation methods: moka, espresso, filtering, and Napolitain way. (Source: http://www.cafeecagliari.it/estrazione/)

coffee powder for preparing an espresso has to be compressed properly, not too much; otherwise it will not allow water percolation (Fig. 4.4). During the process, the water molecules move from the top to the bottom of the coffee layer through a 3D random path.

The problem can be viewed in a more general way considering a liquid poured onto a porous material and the 3D random path which has to go through to reach the bottom; the mathematical modeling of such a problem is the percolation theory.

This theory is one of the best tools to make previsions for systems which randomly grow and has a very broad field of applications, such as an outbreak of flu, a fire spreading in a forest, and polymerization in chemistry. It has been widely applied to study and model the gelation in organic polymers and systems of particles, but it also applies quite well to the sol to gel transition in inorganic and organic-inorganic hybrid systems.

The percolation theory is generally applied using two different models, the *site* and the *bond* percolation.

4.2.2 Site Percolation

In the *site percolation model*, every site is occupied at random with an *occupation probability*, p, or empty with a probability $1 - p$, with p a number between 1 and 0: $0 \leq p \geq 1$.

This model can be easily visualized using, for instance, a square lattice of dimension $L \times L$; each square represents a potential occupation site, which is continuously filled with circles which are randomly distributed within the possible sites (Fig. 4.5).

The probability p in each site is independent of that one of its neighboring. With the increase of the number of the occupied sites, some clusters of different size, s, will form.

Fig. 4.5 Site percolation
in 2D square lattice of
linear size $L = 10$, each site
is occupied with a
probability p. The clusters
in the lattice have
dimensions $s = 7$ (red
circles), $s = 3$ (green), $s = 2$
(blue circles), and $s = 1$,
the isolated sites (yellow
circles)

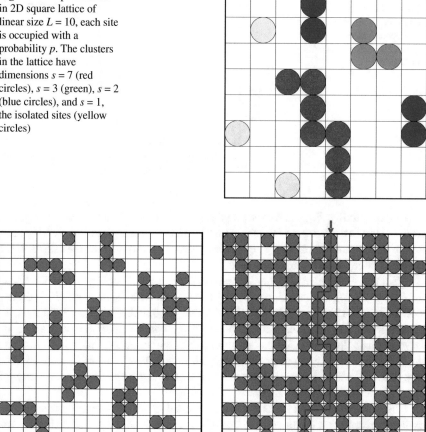

Fig. 4.6 (Left) When only a few clusters (gray circles) occupy the available sites, they are too
small to connect the opposite sides of the square lattice. (Right) If the number of occupied sites is
large, it is possible to connect the top and the bottom of the lattice, for instance, via the red path

When the clusters grow in number and dimension, in fact, it is likely that one of
these clusters may reach a critical dimension and will span from one side to the
other of the grid ($s \to \infty$) (Fig. 4.6). The probability, p, at which an infinite cluster
forms is a critical concentration, *the percolation threshold p_c*.

If the lattice is finite ($L < \infty$), it is clear that if the occupation probability p is
small, the possibility of having a cluster which is able to percolate through the lattice boundaries (for instance, from top to the bottom of the lattice in Fig. 4.6) is also
small. On the other hand, if most of the available sites are occupied (p close to 1),
the percolation of a cluster is almost sure, and several different paths are also
possible.

Fig. 4.7 The probability P_{path} that there is a connecting path between two opposite boundaries sites in a square lattice of dimension L

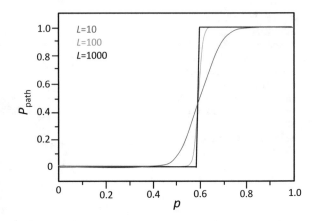

The probability, P_{path}, that exists a path from two boundaries sites (for instance, from the top to the bottom of the square lattice in Fig. 4.6) depends, as we have seen, on the occupation probability p, but clearly also the dimension of the lattice, $L \times L$ (square lattice), is very important [9].

What does it happen with the increase of L? If we look at Fig. 4.7 which shows the change of P_{path} as a function of p, we observe that with the increase of the dimension L, P_{path} tends to approach a critical value, p_c 0.593. For p lower than p_c, the possibility of forming a connection between the two boundaries is virtually zero. Beyond this value, the P_{path} function jumps to 1, which means that it is almost sure the formation of at least one connecting path for $p > p_c$. If $L \to \infty$, P_{path} becomes a step function which jumps directly from zero to 1 at p_c; this describes exactly the physical case of a phase transition.

This is very important to catch because the sol to gel transition is not a thermodynamic event but a continuous transition, so there is not a specific value of temperature, pressure, and volume which describes the transition event which is continuous and defined in statistical terms. It is a series of non-equilibrium aggregation processes that lead to a sol to gel transition.

If we take the p-value as a reference for a transition from a sol to gel, we see that if p is lower than p_c, we still have a sol, while a value of p higher than p_c indicates the formation of a gel:

$$p < p_c \to \text{sol}$$
$$p = p_c \to \text{gelation}$$
$$p > p_c \to \text{gel}$$

In the real world, the percolation threshold will correspond to a change of the macroscopic properties from liquid-like to solid-like when the system stops flowing. The percolation theory can be extended to other more complex cases than a square lattice and closer to the real world, and it is also possible to model the percolation in three or even more mathematical dimensions.

4.2.3 Bond Percolation

The site percolation is very nice to get a first visual step into the theory, but for a gelling system, another model is more appropriate, the *bond percolation*. This model better describes the case of a molecule with several available bonds to react with close and similar molecules. This the typical case of tetraethoxysilane which has four available reacting sites. In the bond percolation, each bond between neighboring lattice sites can be connected (open) with probability p and disconnected (closed) with probability $(1-p)$, and they are assumed to be independent (Fig. 4.8). If we consider two neighboring silicon atoms (each one has four available bonds) within a 3D system, the -Si-O-Si- bond can be formed (occupied site) or not (probability of the event $1-p$). Branching silica molecules can form larger molecules by activating more and more bonds (-Si-O-Si-). This model, at least in this general descriptive approach, appears reasonable to describe a sol to gel transition. At $p = p_c$, a large macromolecule which spans from one side to the other of the vessel would form. The bond percolation, in general, also works quite well for describing many physical transitions, such as isolator-conductors (p_c is the critical concentration which allows the current to flow) or superconductor-conductor in a material.

The values of p_c depend on the lattice dimensionality (1, 2, 3, or higher, as far as computing models are available), lattice type (triangular, 2D square, 3D fcc, 3D bcc, and so on), and coordination number of the bonds which surround a site. We have also to consider that in a sol to gel transition, the branching molecules not only grow by activating more and more bonds but also can form closed structures, such as cages, which is a common finding in sol-gel chemistry. In this case, the percolation model still fails to take into account the different chemical reactions which are behind the sol to gel transition in inorganic systems. Another problem is that the theory does not consider the kinetics of the process; close to the gel point, the cluster growth is likely changing.

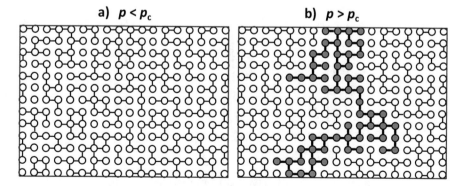

Fig. 4.8 Bond percolation in a 2D lattice, for $p > p_c$ a percolation path from one side to the other forms (black path)

4.2.4   Scaling and Universality

The percolation models have still some other surprises; they allow disclosing some concepts which are *scaling* and *universality*.

As an example, we use bond percolation in a 2D square lattice with p the fraction of connected bonds and $P(p)$ the *percolation probability*, i.e., the probability that a bond is attached to the spanning cluster.

We also introduce two other variables which are the *cluster size*, s, and the *spanning length*, l, which represents the maximum distance between any bond center (or site) in a cluster. The percolation probability function, $P(p)$, obtained by computer simulation in the case of the 2D square lattice is shown in Fig. 4.9. $P(p)$ is zero below p_c but quickly increases, as a function of $p - p_c$ to become $P = p = 1$ in a fully interconnected system. The percolation probability $P(p)$ is the "key function" of the process; beyond p_c, the function indicates how the network grows in volume with the increase of the bond fraction.

In Fig. 4.9 two other curves, which are s_{av} and l_{av}, "exhibit singularities" close to the percolation threshold. They have a very different response in comparison to $P(p)$. At first, we have to notice that they are positive in the region below p_c; in opposition to P and at $p = 0$, where p is also 0, they have a finite value. As soon as the value of p approaches p_c, they show a monotonic growth with a steady slope increase. Close to p_c the divergence of the two functions can be described with power laws which depend on the distance from the threshold:

$$s_{av} \approx \frac{1}{\left(p_c - p\right)^{\gamma}} \tag{4.4}$$

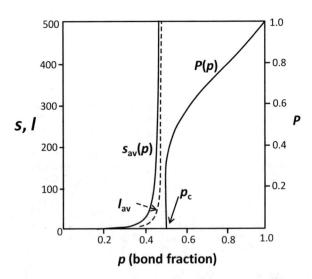

Fig. 4.9 Cluster size, s; spanning length, l; percolation probability, P, as a function of the fraction of the bonds which are formed in the system, p. (Redrawn from Ref. [2])

$$l_{av} \approx \frac{1}{\left(p_c - p\right)^v} \tag{4.5}$$

$$p \rightarrow p_c$$

while the percolation probability is expressed as a function of the bond fraction.

$$P \sim \left(p - p_c\right)^{\beta} \tag{4.6}$$

The interesting point is that even if the percolation threshold, p_c, depends on the lattice type, the exponents of the power law do not depend on the lattice geometry, they are the same for all the lattices of the same dimensionality. They obey a *scaling law*.

The exponents in the previous Eqs. (4.4, 4.5, and 4.6) which describe the behavior of s_{av}, l_{av}, and P functions close to the percolation threshold are always positive not integers numbers which are indicated as *critical exponents*. They are "dimensional invariants," and these exponents rule the power law dependence of the functions in the critical region (close to p_c). For any specific dimensionality, d, which means 2D, 3D, and even higher, as the mathematics has no limits, there is a fixed value of any exponent which is not depending on the short-range organization of the network. This property has been defined universal, to underline the general application in percolation; we have therefore introduced two important concepts which are the scaling laws (critical exponents) and the universality. The percolation threshold, p_c, has not this universality, and the value changes with the lattice type.

Table 4.1 shows the values in two, three, and six dimensions of the critical exponents; the values for the cluster size distribution, $n(s)$, at the percolation threshold when $s \rightarrow \infty$ are also reported.

The last column in Table 4.1 with $d \geq 6$ has a special meaning; in fact with the increase of dimensionality, the percolation theory approaches the classic model. This can be understood if we remind that the coordination number, z_c, increases with the dimensions: $z_c = 2$ ($d = 1$), $z_c = 6$ ($d = 2$), $z_c = 12$ ($d = 3$), $z_c = 24$ ($d = 4$), $z_c = 40$ ($d = 5$), and $z_c = 72$ ($d = 6$); for d the classic theory matches the values of the percolation exponents.

Table 4.1 Critical exponents for near-threshold behavior in percolation theory

Functional form (close to $p = p_c$)	Exponent	Value of exponent in d dimensions		
		2	3	≥ 6
$P(p) \sim (p - p_c)^{\beta}$	β	0.14	0.4	1
$s_{av}(p) \sim (p - p_c)^{-\gamma}$	γ	2.4	1.7	1
$l_{av}(p) \sim (p - p_c)^{-v}$	v	1.35	0.85	1/2
$s \rightarrow \infty : n(s) \sim s^{-\tau}$	τ	2.06	2.2	5/2
$s \rightarrow \infty : n(s) \sim s^{-(1/f)}$	f	1.9	2.6	4

4.3   Percolation and sol to Gel Transition

Percolation, as underlined in the introduction of this chapter, is a general process, and percolation models work quite well to describe different types of critical phenomena [10].

The sol to gel transition in a silica system is an example which fits pretty well with bond percolation. The silicon alkoxide with functionality 4 is the precursor monomer of the process, which in a catalyzed solution forms a reactive monomer. Each silicon alkoxide can form up to four reacted bonds with oxygen bridging two molecules (Fig. 4.10). This is the beginning of the process with the formation of larger molecules that continue to grow. When a macromolecule grows enough to form an extended network that reaches the two sides of the vessel, gelation finally occurs. We have learned this in Chap. 3, but if we look again to this process with the present understanding of percolation, the connection is quite clear.

The number of *reacted bonds* corresponds to the *molecule functionality*; the extent of reactions or *cross-linking probability* is the *fraction of bridging oxygens* (-Si-O-Si-); the ratio between the gel macromolecules and the molecular content (which is given by the sum of the gel macromolecule and the residual sol) gives the gel fraction; and finally the *gel point* is the *percolation threshold*. These analogies are reported in Table 4.2.

The correspondence of the percolation models with a qualitative description of the sol to gel transition in oxide systems is quite clear, but what about the agreement to experimental results? An example can be found in the data reported in Fig. 4.10 where the relative concentrations of the different silica species during the sol to gel transition are shown [11]. The data are relative to an acidic catalyzed sol using tetramethyl

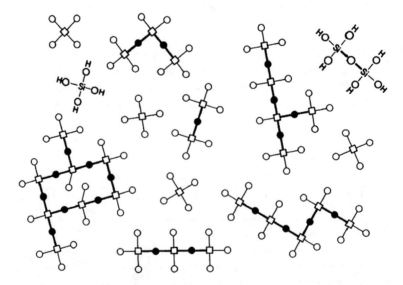

Fig. 4.10 Schematic representation of the condensation reactions in a silica gel. (Reproduced with permission from Ref. [2])

Table 4.2 The percolation model and the analogies with the sol-gel process

Percolation	Gelation
Percolation threshold	Gel point
Connected bond	Bridging chemical bonds
Probability of bond connection	Extent of reaction
Finite cluster	Sol molecule
Mean cluster size, s_{av}	Average molecular weight
Infinite cluster	Gel macromolecule
Percolation probability, P	Gel fraction
Coordination number	Functionality

orthosilicate (TMOS) as silica precursor. The relative concentration of monomeric (M), dimeric (D), trimeric (T), and higher aggregates (P) is shown as a function of time and normalized to the gelation time, t_{gel}. If we compare the experimental curves in Fig. 4.10 with those obtained by computer simulation in Fig. 4.2, which correspond to the change of weight fraction of the different species as a function of reaction degree, we can notice that the qualitative correspondence is pretty good. The experimental data will support the computer simulation; the monomers exhibit a similar trend and quickly decrease in content with the approaching of the gel time, but around 5–10% remain in the system even after gelation (Fig. 4.11).

Other interesting comparisons between experimental data and the models have been made for the average cluster mass and viscosity.

The weight-average cluster mass, M_w, as a function of time for a silica system prepared from TMOS is shown in Fig. 4.12 [6]. M_w slowly increases up to 0.85 t/t_{gel} to diverge close to the gelation point. This trend gives a scaling law for M_w (4.7) [12]:

$$M_w(t) = M_w(0)\left(1 - t/t_{gel}\right)^\gamma \qquad (4.7)$$

In the classic FS theory, $\gamma = 1$; the percolation models give instead a divergence for $\gamma = 1.76$. The experimental data for silica-based sols indicate, however, a stronger divergence (the log-log plot of M_w vs $(1 - t/t_{gel})$ gives a slope of -2.6), with respect both the classic and the percolation models. The experimental data of organic polymers give a γ between 1.5 and 1.8.

To resume, in the different cases, theoretical and experimental, we have the following values for the divergence γ:

$\gamma = 1$	Flory – Stockmayer classic statistical theory
$\gamma = 1.76$	Percolation model
$\gamma = 2.6$	Silica - basedsols
$\gamma = 1.5 \sim 1.8$	Organic polymers

The percolation models also work well to describe the changes in properties observed close to the gel point, such as viscosity. The experimental results produce curves which are similar to $s_{av}(p)$ in Fig. 4.9 which follow a scaling law. The viscosity close to the gel point shows a sharp increase, which corresponds to the divergence of the weight-average molecular weight.

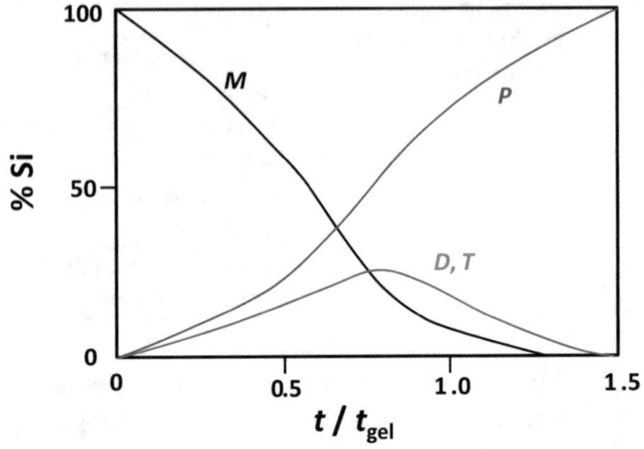

Fig. 4.11 Time evolution (t/t_{gel}) of monomeric (M), dimeric (D), trimeric (T), and higher branched (P) silicon groups. Data obtained by NMR and Raman analysis from a TMOS aging sol. (Redrawn from Fig. 1 in Ref. [4])

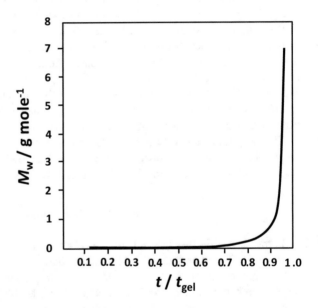

Fig. 4.12 Weight-average molecular weight, M_w, of TMOS sols as a function of relative time t/t_{gel}. (Redrawn from Fig. 6 in Ref. [4])

Figure 4.13 shows the change of viscosity, η, as a function of gelation time for a TMOS sol, such as in the previous examples. In the first stage, the viscosity does not change but quickly increases after $t/t_{gel} = 0.8$. The divergence of the sol near the t_{gel} suggests a power scaling law of the type (4.8):

$$\eta = \left(1 - t / t_{gel}\right)^{s} \tag{4.8}$$

The log-log plot of the curve η vs $(1 - t/t_{gel})$ gives $s = 0.75$.

In the percolation theory, the viscosity near the gelation point becomes (4.9):

$$\eta = \left(1 - p / p_{c}\right)^{s} \tag{4.9}$$

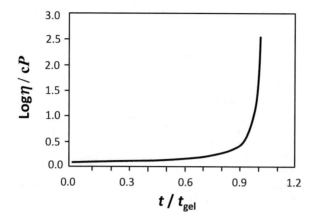

Fig. 4.13 Viscosity, η, of TMOS sols as a function of relative gelation time t/t_{gel}. (Redrawn from Fig. 8 in Ref. [4])

with p the number of bonds in an N-site lattice, with $p = p_c$ at the percolation point. In the systems governed by chemical reactions, p can be replaced by a reaction time close to the percolation threshold to obtain Eq. 4.8. The results are in good agreement with the percolation prediction of $s = 0.7$. On the other hand, the discrepancy between predicted and experimental values of M_w suggests that the models, while working well to give a qualitative description of the process, are not so accurate to take into account the complex chemical reactions involved in the process. The change of reactivity of the active species with the proceeding of hydrolysis and condensation requires a more finely detailed model.

In conclusion, the sol to gel transition is a critical phenomenon which is characterized by the divergence close to the gel point. This behavior can be expressed in terms of power laws of the diverging properties, such as the viscosity, the spanning length, the mean cluster size, and the elastic modulus. Some discrepancies are observed between measured and calculated values of the critical exponents, but, in general, the exponential laws fit well the experimental data and give a good overall description of the sol to gel transition also in inorganic systems.

References

1. Flory PJ (1953) Principles of polymer chemistry. Cornell University Press, New York
2. Kricheldorf HR (2014) Paul J. Flory and the classic theory of percolation. In Polycondensation: history and new results, 35–50
3. Zallen R (1998) The physics of amorphous solids. Wiley & Sons, New York
4. Flory PJ (1941) Molecular size distribution in three dimensional polymers. I. Gelation. J Am Chem Soc 63:3083–3090
5. Flory PJ (1941) Molecular size distribution in three dimensional polymers. III. Tetrafunctional branching units. J Am Chem Soc 63:3096–3100

6. Flory PJ (1942) Constitution of three-dimensional polymers and the theory of gelation. J Phys Chem 46:132–140
7. Pizzi A (2003) Principles of polymer networking and gel theory in thermosetting adhesive formulations. In: Handbook of adhesive technology, revised and expanded. CRC Press, New York, pp 181–192
8. Bailey JK, Macosko CW, Mecartney ML (1990) Modeling the gelation of silicon alkoxides. J Non Cryst Solids 125:208–223
9. Gastner MT, Oborny B (2012) The geometry of percolation fronts in two-dimensional lattices with spatially varying densities. New J Phys 14:103019
10. Cohen-Addad JP (1992) Sol or gel-like behaviour of ideal silica-siloxane mixtures: percolation, approach. Polymer 33:2762
11. Winter R, Hua DW, Song X, Mantulin W, Jonas J (1990) Structural and dynamical properties of the sol-gel transition. J Phys Chem 94:2706–2713
12. Li R, McCoy BJ (2005) Distribution kinetics models for the sol-gel transition critical exponent. Macromolecules 38:6742–6747

Chapter 5
From Silicate Oligomers to Gelation

Abstract The sol to gel transition in inorganic systems is a process highly dependent on the synthesis conditions. The chemistry of silicon alkoxides, in particular, is very difficult to handle because so many different species can form from the very beginning of the process and a small change of the synthesis parameters is immediately reflected on the hydrolysis and condensation stages.

The structure of silica oligomers, in particular, has paramount importance in the process, even if it is generally overlooked when only the final material is considered; the nature of the molecular species has, instead, a strong influence on gelation. Several experimental (Malier, et al. Phys Rev A 46: 959–962, 1992) and theoretical (West, et al. J Non Cryst Solids 121: 51–55, 1990; Tang, et al. J Mater Chem 3: 893–896, 1993) data have shown that silica oligomers contain high concentrations of rings, from three- up to sixfold member rings (Nedelec and Hench J Non Cryst Solids 255: 163–170, 1990). We have also seen that more complex silica structures are likely to be formed as a function of the precursor properties, such as cage-like and cubic silsesquioxane cages. Reactive silica species which have been hydrolyzed can react to each other to form a dimer which has six reactive sites; with the addition of another molecule, a trimer is formed which eventually gives rise to an intramolecular reaction and the formation of a ring. This is a three-member ring which is, however, not very stable. The most common species is, in fact, the four-member ring which is the predominant and more stable structural unit. In this chapter how the nature of the silicate oligomers affect the sol-gel process will be briefly revised.

Keywords Flory-Stockmayer theory · Oligomerization · Effective functionality · Nonrandom cyclization · Gelation

5.1 Failure of the Classic Model

The classic Flory-Stockmayer (FS) theory (*random branching*), as we have seen in the previous chapter, works quite well to predict, at least from a qualitative point of view, gelation and molecular size distribution in organic systems which undergo

© The Author(s), under exclusive license to Springer Nature Switzerland AG 2019
P. Innocenzi, *The Sol-to-Gel Transition*, SpringerBriefs in Materials,
https://doi.org/10.1007/978-3-030-20030-5_5

through a sol-gel transition. The theory, however, does not fit as much as to inorganic gels, which are very far to behave as an ideal system which strictly fulfils the initial restrictive hypothesis of FS theory. Cyclization of silicon alkoxides, because of the complexity of the chemistry involved, plays a very important role from the very beginning of the process, and the system quickly deviates from an ideal model fitting the FS random branching.

To understand in more details this point let us have a look to a specific silica sol that is formed by hydrolysis and condensation of tetramethyl orthosilicate (TMOS, $Si(OCH_3)_4$), dissolved in methanol [5, 6]. The sol-gel reactions can be catalyzed in acid or basic conditions. It is a simple sol-gel system with a silicon alkoxide dissolved in its alcohol; however, when the first oligomers begin to form, they would immediately start to differentiate in length and structure, following different growth paths for basic or acid conditions.

It is possible to obtain an accurate distribution of the different silicon species by capillary gas chromatography. Table 5.1 lists the different structures observed in the sol, from mono to hexasilicates.

Table 5.1 Observed silica structures, from mono- to hexasilicates, in a TMOS sol

Formula	Structure	Label
$Si(OCH_3)_4$		ℓ-Si_1
$[Si_2O](OCH_3)_6$		ℓ-Si_2
$[Si_3O_2](OCH_3)_8$		ℓ-Si_3
$[Si_4O_4](OCH_3)_8$		c-Si_4
$[Si_4O_3](OCH_3)_{10}$		b-Si_4
		ℓ-Si_4
$[Si_5O_5](OCH_3)_{10}$		c-Si_5
$[Si_5O_4](OCH_3)_{12}$		b-Si_5
		ℓ-Si_5
$[Si_6O_6](OCH_3)_{12}$		c-Si_6
$[Si_6O_5](OCH_3)_{14}$		b-Si_6
		ℓ-Si_6

Redrawn from Ref. [5]

Three types of silicon species are observed to form:

$$Linear \quad 1-\text{Si}_n$$
$$Branched \quad b-\text{Si}_n$$
$$Cyclic \quad c-\text{Si}_n$$

while the distribution of the species, formed under basic and acid conditions, is listed in Table 5.2.

These different silica structures contain Si centers which are connected to a maximum of four adjacent silicon atoms by bridging oxygens. They form Q^n different Si environments with $n = 0-4$. The data in Table 5.2 give two important indications: (1) most of the silica species formed under acid conditions are linear and (2) the distribution of species which is obtained in acid conditions is quite narrow, with 92% of silica which falls within the mono- to hexasilicate size range. In the case of basic sols, the distribution is broader, and only 66% of the silicon in the system is contained in the same size range. The molecular size distribution of the different species in Table 5.2 is visualized in the two curves in Fig. 5.1 for basic and acid sols. The points are the experimental data; the lines are a guide for the eyes.

The previous data can be now compared to the idealized sol-gel silica polycondensation described by the Flory-Stockmayer (FS) theory. The theory, as we have seen in the previous chapter, describes the polymerization of monomers with a functionality, f, but with some restrictive assumptions, such as the exclusion of cyclic species and considering that all the monomers do not change their reactivity during the process. Following the FS theory, the molecular size distribution is given by (5.1):

$$\%\text{Si}(n) = 100 \times \frac{(fn-n)!\,f}{(fn-2n+2)!(n-1)!}\left(\frac{2r}{f}\right)^{n-1} \times \left(\frac{f-2r}{f}\right)^{fn-2n+2} \tag{5.1}$$

Table 5.2 Distribution of the different silicon species, from mono- to hexasilicates, in a TMOS sol catalyzed in basic and acid conditions [5]

Silicon species	Mole % Si	
	Acid conditions	Basic conditions
ℓ-Si$_1$	21.8	38.6
ℓ-Si$_2$	22.4	16.0
ℓ-Si$_3$	18.7	6.6
c-Si$_4$	0.9	0.3
b-Si$_4$	2.2	0.5
ℓ-Si$_4$	10.9	2.3
c-Si$_5$	1.6	0.3
b-Si$_5$	2.9	0.4
ℓ-Si$_5$	5.3	0.7
c-Si$_6$	1.1	< 0.05
b-Si$_6$	2.4	–
ℓ-Si$_6$	2.1	< 0.10

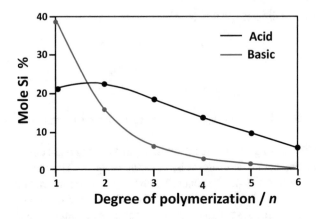

Fig. 5.1 Molecular size distribution of silicate species in an acid (black line) or basic (red line) catalyzed TMOS sol. The lines are a guide for eyes; the points are the experimental data. (Redrawn from Ref. [5])

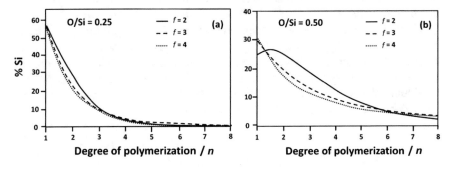

Fig. 5.2 Flory-Stockmayer molecular size distribution calculated using the Eq. 5.1 for **(a)** O/Si = 0.25 and **(b)** O/Si = 0.5. (Redrawn from Ref. [5])

with %Si (n) the percentage of silicon atoms which are in the $[Si_nO_{n-1}](OCH_3)_{2n+2}$ molecules, f the monomer functionality, and r the bridging oxygen to silicon ratio, O/Si, for the system. The variable r gives the extent of the polycondensation of the system.

Fig. 5.2 shows the results of the theoretical calculations for a system at an early stage of reaction (low O/Si ratio, 0.25) (Fig. 5.2a) and with a higher degree of condensation (O/Si = 0.5) (Fig. 5.2b). The case $f = 2$ can be considered as an idealized linear polymerization (only Q^0, Q^1, and Q^2 species are allowed), while $f = 3$ and $f = 4$ represent branched polymerization with $Q^0 - Q^3$ and $Q^0 - Q^4$ species allowed, respectively.

At the beginning of the sol-gel process (O/Si = 0.25), the distribution of silicon species as a function of the polymerization degree shows no significant changes. The curves for $f = 2$, 3, or 4 show a similar trend.

If we compare the theoretical curves in Fig. 5.2 with the experimental data in Fig. 5.1, we see that a good correspondence exists between the species observed in acid conditions, which we know are mostly linear, and the $f = 2$ curve of the Flory-Stockmayer theory. The data for basic synthesis are well correspondent also to the branched polymerization ($f = 3, f = 4$). This is, however, a qualitative corresponding, but as we know from the previous chapter, a strong discrepancy is, instead, observed with the experimental data regarding the total amount of bonds which are necessary to observe gelation. FS theory gives a substantial underestimation because of the restrictive assumptions.

5.2 Silica Cyclization

The percolation theory considers the possibility of the random formation of closed loops of different sizes. The specificity of sol-gel silica chemistry makes the closure of rings an event which is instead kinetically favored with respect to the formation of chain-like structures. The typical example is the acid-catalyzed hydrolysis and condensation of TEOS. This peculiar ring formation, defined as *nonrandom cyclization*, is likely the main cause of the experimental mismatch with the FS theory. The different reactivity of the silica species which form from the hydrolysis of the precursor through the condensation is also another important cause of the deviation from the classic and percolation models.

In an ideal sol-gel silica system following the classic model, a transition upon conversion of around 33% of the bonds is observed, but this does not correspond to the experimental observation [7]. The fraction of bond conversion, α, can be followed by measuring the fraction of silicon atoms with n siloxanes bridges by ^{29}Si NMR. In the case of a silicon tetrafunctional alkoxide (5.2):

$$\alpha = \frac{1}{4}\sum_{i=0}^{4} i\left(Q^i\right) \qquad (5.2)$$

with Q^0, Q^1, Q^2, Q^3, and Q^4, which corresponds to silicon atoms with 0, 1, 2, 3, and 4 siloxane bridges, respectively.

Figure 5.3 shows the conversion rates in silica sols using TEOS as precursors [7] at the beginning of the reaction stage [8] with different acid and water concentrations. The bond conversion shows that after 4 hours of reaction, up to 50% of the available reactive sites have been converted to bridging silica. In this first reaction stage, the bond conversion is depending on the water and acid concentrations. This dependence is an indirect indication of the simultaneous presence of different intermediate oligomeric silica species, whose concentration is controlled by the synthesis conditions. Branching and cyclization are forming different silica oligomers, which affect the gelation process through their specific reactivity.

If the experiment is extended up to the gelation time, it is observed something very interesting; beyond the conversion of 60–70%, all the systems converge to a

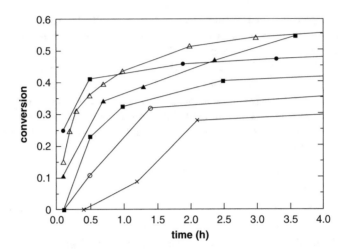

Fig. 5.3 Hydrolysis and condensation of TEOS: early in the reaction, conversion rate varies with acid and water concentrations. Compositions (TEOS/EtOH/H$_2$O/HCl) in mole/liter: (•) 2.02 / 8.11 / 4.04 / 0.012; (■) 2.02 / 8.11 / 4.04 / 0.003; (x) 2.02 / 8.11 / 4.04 / 0.001; (▲) 1.89 / 7.56 / 7.54 10.006; (O) 1.89/ 7.56 / 7.54 / 0.002; (Δ) 1.32 / 7.94 / 13.24 / 0.001. (Reproduced with permission from Ref. [7])

plateau with the limit value of $\alpha = 83\%$ (Fig. 5.4). Different researchers have measured a similar threshold value in various experimental conditions, and it is considered a reliable general conversion value for gelation of silica systems. At the molecular level, the polycondensation of silica depends on *nonrandom cyclization* which delays the bond conversion from $\alpha = 33\%$, expected from the classic model to $\alpha = 83\%$ which is the experimental value.

The conversion rate of 83% is also higher than the value predicted by the percolation theory which allows statistical branching and cyclization. However, even if *random cyclization* and the *effective functionality, f,* during the reaction are taken into account, the gel conversion rate only slightly increases (Table 5.3) [7]. The *effective functionality* of a monomeric unit corresponds to the average number of silanols and silica bridging species and evolves with the reaction time. Table 5.3 shows a comparison of gel conversion predictions in the case of the percolation and classic theories. The *effective functionality* affects the conversion in the same way, and the value remains much lower than the experimental results.

A detailed correlation between the degree of silica condensation during sol to gel transition and the relative content of silica bridges can be obtained by following the change of Q^i species as a function of gelation time (Fig. 5.5) [8].

With the proceeding of hydrolysis and condensation, the unreacted monomers (Q^0) quickly decrease with the rise of Q^1, Q^2, and Q^3 species which finally decrease with a different rate during the formation of a fully condensed silica network and the consequent rise of Q^4. The condensation degree increases with the gelation time and reaches a value of 0.81, which is again an experimental evidence of the failure of a random branching model.

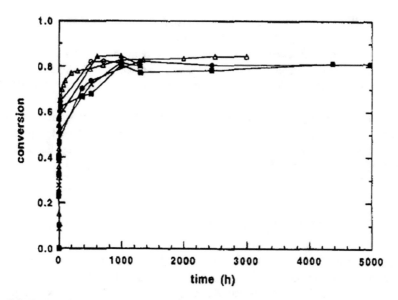

Fig. 5.4 Hydrolysis and condensation of TEOS: later in the reaction, conversions all slow down and converge to 83%. Gel times: (•) 5064 h; (■) 4392 h; (x) ~1700 h; (▲) ~1500 h; (○) 1020 h; (△) 1200 h. (Reproduced with permission from Ref. [7])

Table 5.3 Gel conversion from random growth as a function of effective functionality for classic (FS) and percolation theories [7]

Effective functionality	Gel conversion from random growth	
	Percolation theory (branching and cyclization)	Classic (FS) theory (branching)
3	0.5	0.5
4	0.375	0.333
5	0.3	0.25
6	0.25	0.2

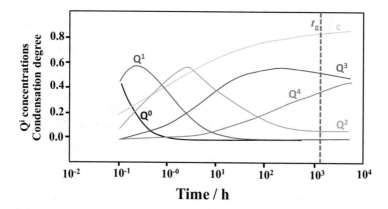

Fig. 5.5 Time evolution of the concentration of the different Q^i species and the degree of condensation for an acid-catalyzed TEOS sol prepared with TEOS/EtOH/H_2O = 1:6:10 molar ratios and pH 2.5. (Redrawn from Fig. 3 in Ref. [8])

If the FS model is used to make a prevision of the gelation time, the random branching model also gives a predictive value which is far from the experimental one. The difference between the theoretical bond conversion (33%) and the measured value of 83% is reflected in the discrepancy between the predicted time of 4 h and the measured gelation time of 35 days in an acid-catalyzed silicon alkoxide system [9].

The linear chain growth with random branching [10, 11] in most of the cases is an unrealistic model to describe the sol oligomeric structure and the sol to gel transition. Chain growth appears instead of a competing process with cyclization reactions in the formation of silica building blocks [7].

5.3   The Oligomerization Pathway

The comparison between theoretical models and experimental values for the sol to gel transition underlines the importance of controlling the oligomerization pathways of silica [12, 13]. In the literature, several detailed studies of the silica species which form during the sol-gel process have been reported [14]. A detailed study has been obtained by slowing down the reaction kinetics in under stoichiometric and strongly acidic conditions for TEOS [15] and TMOS [16] systems. The results in the case of a TEOS sol prepared with a water-alkoxide ratio of 0.7 are shown in Fig. 5.6. The monomeric species quickly decrease in accordance with data in Fig. 5.5; the same trend is also shown by dimers even if they are not completely consumed. At longer reaction times, chains, rings, and branched silica oligomers grow at expenses of smaller units, basically monomers and dimers that give rise to cyclization or branched oligomers.

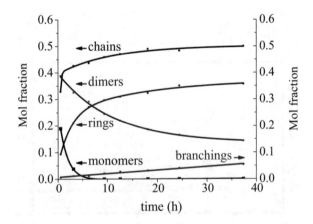

Fig. 5.6 Evolution of mole fractions of monomer and different types of oligomers in a sample prepared using an H_2O/TEOS = 0.7 (derived from Q^0, Q^1, and Q^2 signals in ^{29}Si NMR) and a fraction of branching expressed as the fraction of Q^3 in the totality of silicon atoms. Lines are meant to guide the eye. (Reproduced with permission from Ref. [15])

The experimental data also allow going more in detail about the changes with the reaction time of the cyclic species and the evolution of rings of different sizes, from threefold up to sixfold rings (Fig. 5.7). The more unstable threefold rings quickly disappear with the proceeding of the reaction, while the total content of rings of larger dimension shows the tendency to increase with time, with a prevalence of formation of fourfold rings which are the most common cyclic species [17].

At this point it is possible to get a more general overview of the silica oligomers formation pathway; the sol is a mixture of different oligomeric species which grow at expenses of monomers and dimers. The reaction of smaller units tends to form, through intermolecular reactions, branched species which eventually grow to span as an extended network.

There is another interesting question to answer, how do the ring structures form? Two main routes are possible: the first one is intramolecular cyclization; this means that as shown in Fig. 2.10, a linear trimer, tetramer, pentamer, or hexamer can react to form a closed intramolecular loop (cyclization of chains). On the other hand, it is also possible forming a ring by the reaction of dimers and trimers (cyclodimerization). Between these two scenarios, the second one is more likely because an increase of tetramers accompanies the increase of fourfold rings population. This increase is in contradiction with the possibility of intramolecular reaction to form fourfold rings which should, instead, lead to a consumption of silica tetramers.

The scheme of TEOS oligomerization in a sol prepared in acid and substoichiometric amount of water is shown in Fig. 5.8.

It should always be taken into account that the chemistry and the oligomerization pathways can quickly change as a function of synthesis parameters. If we use TMOS instead of TEOS, significant differences are already observed [18]; in TEOS systems the rings form at an early stage of reaction via cyclodimerization, while in TMOS rings appear only in a later stage of reaction [16].

Figure 5.9 shows the evolution of the silicate chain length distribution for acid-catalyzed TMOS system ($r = 0.9$). In the first hours, the content of dimers (2Si) quickly decreases, from ~30 mol% after 30 min to ~12 mol% after 20 h. The 3Si and 4Si chains disappear to form longer oligomers with five and more Si atoms.

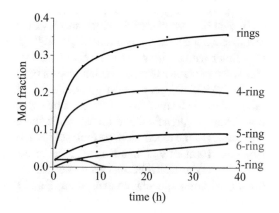

Fig. 5.7 Evolution of the mole fractions of cyclic oligomers and of the sum of these ring structures in a sample prepared using an $H_2O/TEOS = 0.7$. (Reproduced with permission from Ref. [15])

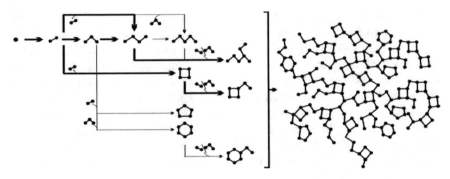

Fig. 5.8 Oligomerization scheme of TEOS in an acid medium and substoichiometric quantities of water. (Reproduced with permission from Ref. [15]). The thickness of the arrow indicates the significance of the reaction pathway

Fig. 5.9 Evolution of the molar distribution (%) of silicate chain lengths in a sample prepared using a $H_2O/TEOS = 0.9$. (Redrawn with permission from Ref. [16]). Experimental data points were left out for clarity

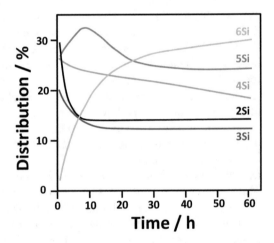

The amount of 5Si chains show a maximum of around 10 h to decrease at longer reactions times to give rise to species with six and more Si atoms.

In general TMOS-derived silica species favor the formation of branched oligomers via a condensation reaction of Q^1 and Q^2 groups. An example is the reaction 1 of a dimer with a trimer to form a Si_5 oligomer and the reaction 2, between a tetramer and a pentamer whose reaction gives a branched Si_9 oligomer (Fig. 5.10).

The TMOS sol-gel chemistry is quite different from that of TEOS, and this is reflected in the formation of different silica species upon hydrolysis and condensation. They are branched and linear, while the cyclic species are observed only at later stages of the reaction. If we look at Fig. 5.11, it is possible to observe that rings structures form only at an advanced level of condensation. In the first silica structures, from A to D, cyclic species do not form; they rise instead only with the proceeding of intramolecular condensation, E and F stages in the figure.

Fig. 5.10 Example of branched chain formation via reaction of Q^1 and Q^2 groups which form a pentamer Si_5 and a Si_9 oligomer

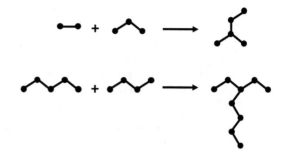

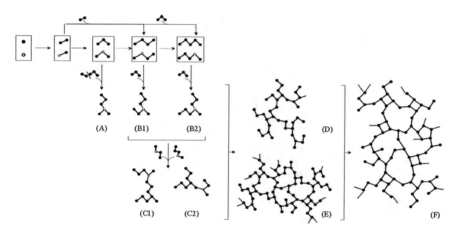

Fig. 5.11 Acid-catalyzed silica oligomerization scheme departing from TMOS (Fully methoxide terminated and hydrolyzed silicon atoms are presented by solid (●) and open (○) dots, respectively)

References

1. Malier L, Boilot JP, Chaput F, Devreux F (1992) Nuclear-magnetic-resonance study of silica gelation kinetics. Phys Rev A 46:959–962
2. West JK, Zhu BF, Cheng YC, Hench LL (1990) Quantum chemistry of sol-gel silica clusters. J Non Cryst Solids 121:51–55
3. Tang A, Xu R, Li S, An Y (1993) Characterisation of polymeric reaction in silicic acid solution: intramolecular cyclization. J Mater Chem 3:893–896
4. Nedelec JM, Hench LL (1990) Ab initio molecular orbital calculations on silica rings. J Non Cryst Solids 255:163–170
5. Flory PJ (1941) Molecular size distribution in three dimensional polymers I. Gelation. J Am Chem Soc 63:3083
6. Yamane M, Inoue S, Yasumori A (1984) Sol-gel transition in the hydrolysis of silicon methoxide. J Non Cryst Solids 63:13–21
7. Ng LV, Thompson P, Sanchez J, Macosko CW, McCormick AV (1995) Formation of cagelike intermediates from nonrandom cyclization during acid-catalyzed sol-gel polymerization of tetraethyl orthosilicate. Macromolecules 28:6471–6476

8. Devreux F, Boilot JP, Chaput F, Lecomte A (1990) Sol-gel condensation of rapidly hydro-lyzed silicon alkoxides: a joint ^{29}Si NMR and small angle X-ray scattering. Phys Rev A 41:6901–6909
9. Bailey JK, Macosko CW, Mecartney ML (1990) Modeling of gelation of silicon alkoxides. J Non Cryst Solids 125:208–223
10. Gnado J, Dhamelincourt P, Pelegris C, Traisnel M, Mayot AL (1996) Raman spectra of oligo-meric species obtained by tetraethoxysilane hydrolysis-polycondensation process. J Non Cryst Solids 208:247–258
11. Mulder C, Vanleeuwenstienstra G, Woerdman JP (1986) Chain-like structure of ultra-low den-sity SiO_2 sol-gel glass observed by TEM. J Non Cryst Solids 82:148–153
12. Trinh TT, Jansen APJ, van Santen RA, Meijer EJ (2009) Role of water in silica oligomeriza-tion. J Phys Chem Lett 113:2647–2652
13. Henschel H, Schneider AM, Prosenc MH (2010) Initial steps of the sol-gel process: modeling silicate condensation in basic medium. Chem Mater 22:5105–5111
14. Brunet F, Cabane B (1993) Populations of oligomers in sol-gel condensation. J Non Cryst Solids 163:211–225
15. Depla A, Lesthaeghe D, van Erp TS, Aerts A, Houthoofd K, Fan F, Li C, Van Speybroeck V, Waroquier M, Kirschhock CEA, Martens JA (2011) ^{29}Si NMR and UV-Raman investigation of initial oligomerization reaction pathways in acid-catalyzed silica sol-gel chemistry. J Phys Chem C 115:3562–3571
16. Depla A, Verheyen E, Veyfeyken A, Van Houteghem M, Houthoofd K, Van Speybroeck V, Waroquier M, Kirschhock CEA, Martens JA (2011) UV-Raman and ^{29}Si NMR spectroscopy investigation of the nature of silicate oligomers formed by acid catalyzed hydrolysis and poly-condensation of tetramethylorthosilicate. J Phys Chem C 115:11077–11088
17. Trinh TT, Jansen APJ, van Santen RA (2006) Mechanism of oligomerization reactions of sil-ica. J Phys Chem B 110:23099–23106
18. Kelts LW, Armstrong NJ (1989) A silicon-29 NMR study of the structural intermediates in low pH sol-gel reactions. J Mater Res 4:423–433

Chapter 6
Measuring the Sol to Gel Transition

Abstract This chapter is dedicated to the methods and techniques which have been used to assess and measure the sol to gel transition and the gel point. The rheological methods have been the first to be applied even if the change in viscosity is not a reliable way to define the gel point. The viscosity gives, however, direct information on the change of the system during the sol to gel transition. A better definition of the gel point has been obtained using dynamic viscoelastic experiments and the storage and loss moduli. The Winter and Chambon relation has introduced the possibility to achieve a precise identification of the gel point. Another technique which has developed is the time-resolved dynamic light scattering which allows the identification of the gel point through different methods.

Keywords Viscosity · Rheological properties · Viscoelasticity · Loss modulus: Storage modulus

6.1 Measuring the Gel Point and the Gel Time

In the first chapter, the definition of sol, gel, and sol to gel transition has been introduced; the main issue which remains to be discussed is how to measure the gel point in the different systems. This is not a trivial aspect because only indirect measures are possible and no latent heat is released during the transition. The evaluation of macroscopic properties, such as viscosity and shear stress, gives a good indication of the transition between the two states, the sol and the gel, but a clear measure of the exact gel point remains highly challenging. This is also reasonable because we know now that a gel is formed through a continuous process. The gelation point is reached when the largest spanning cluster forms a structure which is completely interconnected from one side to the other of the container, but there is no physical discontinuity to mark the process. The models to describe the sol to gel transition have shown, however, that close to the gel point, a clear divergence of some properties should be expected. It is exactly this divergence which is generally measured and is used as an indicative measure of the transition.

© The Author(s), under exclusive license to Springer Nature Switzerland AG 2019
P. Innocenzi, *The Sol-to-Gel Transition*, SpringerBriefs in Materials,
https://doi.org/10.1007/978-3-030-20030-5_6

The simplest way to observe gelation is putting a sol in a vessel and observing when the content does not drop out when "the meniscus of a sol in a container no longer remains horizontal when the container is tilted" [1]. This is an easy approach for getting a qualitative idea of the gelation but is not very precise and leaves some room for a subjective evaluation of the gel time. The things are even much more complicated if we try to measure the gel point in a film, which is a fast evaporating system that requires a special approach.

Monitoring the rheological properties, such as viscosity [2] and viscoelasticity [3], during the sol to gel transition in a bulk gel is one of the most common methods to evaluate the gelation time, t_{gel}. This is not the only information which is possible to obtain from rheological measures; in fact, if the viscous or mechanical responses are correlated with the synthesis parameters, for instance, composition, pH, and concentration, some details of the microstructure can also be deduced.

6.2   Measuring the Gel Point Through the Rheological Properties

6.2.1   Viscosity

The divergence of a macroscopic property, such as the viscosity, η, is an indication of the drastic changes that occur close to the gel point during the transition from the sol phase (see Fig. 4.13). As we know, the sol becomes macroscopically rigid because of the formation of an interconnected giant cluster. This macroscopic change in the rheological properties is not accompanied, however, by a similar drastic variation at the molecular level. No reduction of the mobility of any molecule in the liquid phase is observed at the gel point because of the open structure of the gel. This is an important point to stress; the rigidity at macroscopic level upon gelation is not conversely followed by a similar rigidity at the molecular level at least in submicron species dispersed in the sol which represents the liquid phase of the gel. In the next chapter, we will see more in detail the changes in the microenvironments within the gel phase during the sol to gel transition.

Viscosity as a macroscopic property diverges, as we have seen in the previous chapter, close to the gel point; the general trend is well represented in Fig. 6.1, which shows the viscosity of a ZrO_2 sol, a function of the reaction time [4]. Before approaching the gelation time, the viscosity only slightly increases but close to the sol to gel transition point abruptly rises from 2 to 18.4 mPa·s in a few minutes. The gelation point shifts to longer times with decrease of sol concentration, but the viscosity curves do not change, in accordance with the percolation theory.

Measuring the viscosity during the sol to gel transition also gives some structural information, because the reduced viscosity, η_{sp}/C, (with C the polymer concentration), obeys two different laws if the species in the sol are linear [5] or spherical [6]. Different preparation conditions, in particular acid or basic catalysis, give different

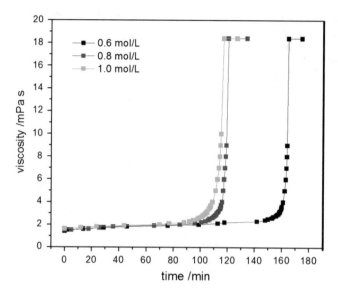

Fig. 6.1 Viscosity as a function of reaction time for ZrO_2 sols with different concentrations. (Reproduced with permission from Ref. [4])

types of silica condensed species. The structure of these clusters affects the viscosity of the material whose response can be described in terms of the following equations:

$$\text{Linear polymers} \rightarrow \text{Huggins equation} \rightarrow \eta_{sp}/C = [\eta] + K[\eta]^2 C \quad (6.1)$$

$$\text{Spherical polymers} \rightarrow \text{Einstein relations} \rightarrow \eta_{sp}/C = K/\rho \quad (6.2)$$

with η_{sp}, the specific viscosity defined as $\eta_{sp} = \eta_{rel} - 1$, and η_{rel}, the relative viscosity; $[\eta]$ the intrinsic viscosity, defined at infinite polymer dilution and obtained by extrapolation of η_{sp}/C vs C zero concentration of the polymer; K, a proportionality constant; and ρ, the polymer density.

The two equations describe the change in viscosity induced by the different structure of the sol; if the species are spherical, there is no dependence on the concentration and particle size (Eq. 6.2). These relations have been applied by Sakka et al. [7, 8] to acid- (HCl) and basic (HNO_4)-catalyzed silica (TEOS) systems and have confirmed that chain-like or linear polymers are preferentially formed via acid catalysis, while spherical species are given by basic catalysis [9, 10].

Figure 6.2 shows the intrinsic viscosity as a function of t/t_g in acid and in basic TEOS-catalyzed sol. In the acid sol, $[\eta]$ shows a gradual increase with time, while in the basic sol it remains constant at low values up to $t/t_g \sim 0.6$ when a fast increase of $[\eta]$ close to the gelation point is observed. In the same experiments, it has also been correlated the change of intrinsic viscosity with the concentration of the oxide

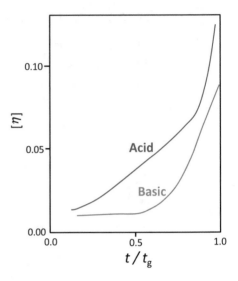

Fig. 6.2 Intrinsic viscosity, [η], as a function of t/t_g ($T = 25$ °C) in an acid or basic TEOS-catalyzed sol. (Redrawn from Ref. [8])

and has been observed that only the acid sol shows a dependence of [η] with the concentration.

In general, an increase of intrinsic viscosity as a function of reaction time is an indication of the enhancement of polycondensation in the sol. The response of the intrinsic viscosity function gives a direct indication of structural differences between the two sols. The intrinsic viscosity in an acid sol is always higher, at the same oxide concentration than in a basic catalyzed sol. In general, solutions which contain linear polymers are more viscous than those containing spherical polymers. The independence of [η] on the concentration also gives a good indication that the different response in the basic catalyzed sol can be attributed to silica spherical-shaped particles which aggregate to form a 3D structure. A strong limitation of a viscosity measured in steady shear flow is the effect of the shear rate.

An example is reported in Fig. 6.3, where the change in viscosity as a function of aging time ($T = 25$ °C) for an acid-catalyzed sol (H_2O/TEOS = 2, EtOH/TEOS = 5.8, and HNO_3/TEOS = 0.1) is shown [12]. The viscosity has been measured at three different shear rates, 3, 300, and 2700 s^{-1}. Close to the gel point (around 1176 hours), the viscosity depends on the shear rate, and the system does not behave anymore as a Newtonian fluid, which should be independent on the shear rate. The sol transition from a Newtonian to a shear thinning (the viscosity increases with the rate of shear strain) and to a thixotropic system (the viscosity decreases with time when the fluid is shaken, agitated, sheared, or stressed. The longer the fluid undergoes shear stress, the lower is its viscosity) reflects the formation of a growing network. This is a general finding, at least for silica sols, with aging the system exhibits a transition in the flow behavior when is approaching the gel point. These transitions are clearly correlated with the structural changes of the sol. Even if this behavior is generally observed in silica sols, some differences can be observed in acid and catalyzed systems if the viscosity data are plotted on a dimensionless time scale. Figure 6.4 shows

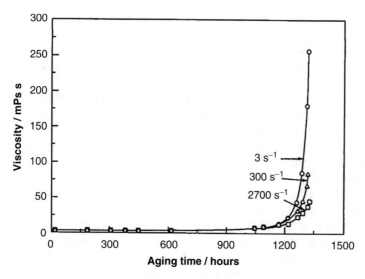

Fig. 6.3 Plots of viscosity at different shear rates (3, 300 and 2700 s⁻¹) as a function of the aging time (hours). The silica sol has been prepared with the following molar ratios: $H_2O/TEOS = 2$, $EtOH/TEOS = 5.8$, and $HNO_3/TEOS = 0.1$. (Reproduced with permission from Ref. [12])

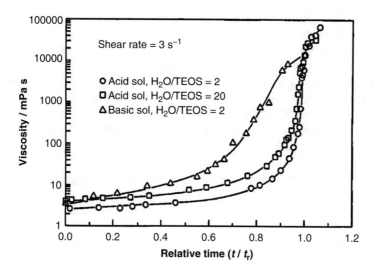

Fig. 6.4 Viscosity (measured at 3 s⁻¹) shear rate changes as a function of the relative time (real time/arbitrary time) in the case of three different sols, an acid sol, $H_2O/TEOS = 2$; a diluted acid sol, $H_2O/TEOS = 20$; and a diluted basic sol, $H_2O/TEOS = 20$. (Reproduced with permission from Ref. [12])

how the viscosity (measured at 3 s⁻¹ shear rate) changes as a function of the relative time (real time/arbitrary time) in the case of three different sols, an acid sol, $H_2O/TEOS = 2$; a diluted acid sol, $H_2O/TEOS = 20$; and a diluted basic sol, $H_2O/TEOS = 20$. The acid sols show a sudden transition in viscosity, with a little difference for the diluted sol which has a small delay in the transition. This viscosity change is more gradual in the basic sol; this should be an indication that aggregates of larger dimensions form at a relatively early stage. The measure does not give a quantitative evaluation of the growth for basic and acid systems but supports what we know about the dependence of the gel structures upon the hydrolysis and condensation parameters.

The measure of viscosity as a function of aging time using rotational methods has strong limitations, as we have seen, that is, shear rate dependent. The experimental data obtained by this method have been found to fit in general very well the power law. Figure 6.5 shows the normalized viscosity as a function of the reaction time for a hybrid organic-inorganic system composed by ZnO nanoparticles, a prehydrolized silica precursor (TES40), and polydimethylsiloxane (PDMS) [11]. The data well agree with the viscosity power law that we have described in Chap. 4:

$$\frac{\eta(t)}{\eta_0} = \left(1 - \frac{t}{t_{gel}}\right)^{-s} \tag{6.3}$$

with η_0 the viscosity at $t = 0$ and t_{gel} the gelation time ($t_{gel} = 30.2 \pm 0.1$ hours). The value obtained for the critical exponent, s, for the system in Fig. 6.5 is 0.72 ± 0.2 which is close to other experimental value measured in other silica system and also close to the 0.7 value predicted by the percolation theory.

The viscosity is a good indicator of the macroscopic changes which are observed during the sol to gel transition; it also depends on the size and shape of the molecular species in the sol. We could be tempted to associate a specific value of viscosity to the gelation time, but the peculiar preparation conditions affect so much the

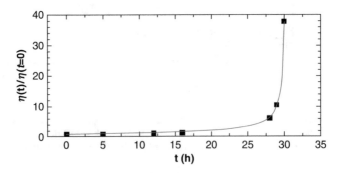

Fig. 6.5 Normalized viscosity measured as a function of time (squares) and power law fit, $(1 - (ŭ61;/ŭ61;_{ŭ54;ŭ52;ŭ59;}))^{-ŭ60;}$, to the experimental data, where t_{gel} and s are fitting parameters (continuous line). (Reproduced with permission from Ref. [11])

change of viscosity during the process that large differences can be observed even in systems of similar composition. Another limitation is that the measurement of viscosity as a function of time is generally done in conditions of constant shear stress or shear rate, but, as we have just seen, the viscosity changes close to the gel point depend on the shear rate used for the measure. A possible solution to overcome this problem is using dynamic methods, such as oscillatory shear measurements.

6.2.2 Viscoelastic Experiments for Determination of the Gel Point

One of the most precise methods of probing the sol to gel transition is performing a *dynamic viscoelastic experiment*. This technique has been developed for organic polymers but has also been applied to inorganic gels in bulk [12–14] and colloidal suspensions [15]. Small amplitude oscillatory shear measurements are particularly suitable to monitor the continuous changes during the sol to gel transition.

The technique measures the viscoelastic response of the gel as a function of shear rate expressed by the *complex shear modulus G*, which is the stress response to a harmonic strain excitation (6.4):

$$G = G'(\omega) + iG''(\omega) \tag{6.4}$$

with ω the frequency of oscillation of the probe and:

$$G' = \textit{storage modulus} \rightarrow \text{elastic component}$$
$$G'' = \textit{loss modulus} \rightarrow \text{viscous component}$$

The two components, the elastic and the viscous one, are expressed in term of *loss tangent*, tang δ, which gives a relative measure of the viscous energy losses with respect to the energy stored in the system (6.5):

$$\text{tangd}\delta = G'/G'' \tag{6.5}$$

An example of application to silica sol-gel systems (TEOS) is shown in Fig. 6.6 where the storage and loss moduli, G' and G'', are plotted as a function of time. The storage modulus shows a sharp increase which can be correlated to the formation of an interconnected network which can develop an elastic mechanical response [16].

If we know the curves of G' and G'' as a function of reaction time, we can trace tang δ, which is given by the ratio between G' and G'' (Eq. 6.5). The change of tang δ with aging time is shown in Fig. 6.7. As soon as it is approaching the gelation time, a small increase of tang δ, which is due to the formation of a more interconnected structure within the gel, is observed. This is because in the transition zone, the elastic component has a faster response than the viscous one. The loss tangent reaches a maximum whose position corresponds to the gelation time, t_g.

Fig. 6.6 The storage
modulus and the loss
modulus as a function of
sol aging time in a silica
sol (H_2O/TEOS = 2, EtOH/
TEOS = 5.8, and HNO_3/
TEOS = 0.1). (Redrawn
from Ref. [12]. The
experimental points have
been omitted for sake of
clarity

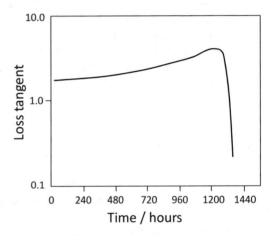

Fig. 6.7 Loss tang δ as a
function of aging time
(H_2O/TEOS = 2, EtOH/
TEOS = 5.8, and HNO_3/
TEOS = 0.1). (Redrawn
from Ref. [12]). The
experimental points have
been omitted for the sake
of clarity

This is, however, not the most precise way of measuring the gel time. Using the experimental measures of G' and G'', another criterion can be applied for identifying the gelation time; it is the Winter and Chambon relation [17–19]. They have observed that at the gel point, the values of the storage and loss moduli are almost the same (6.6):

$$G'(\omega) \sim G''(\omega) \sim \omega^{\Delta} \qquad (6.6)$$

with Δ the relaxation exponent and ω the frequency of measurement. The ratio between the moduli is also expressed by (6.7):

$$\left[G'(\omega) / G''(\omega) \right]_{t=t_g} = \tan(\Delta\pi / 2) \qquad (6.7)$$

The time when the G''/G' ratio is independent on the frequency defines, therefore, the gelation time. The crossing point of the storage and loss moduli, G' and G'',

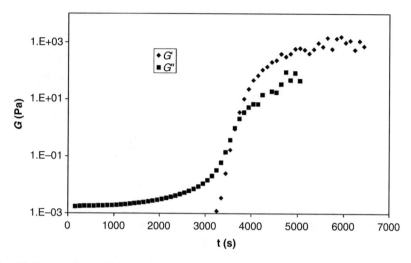

Fig. 6.8 Time evolution of the storage G' and loss G'' moduli for initial monomer concentration TMOS = 1.2 mol L^{-1}, hydrolysis molar ratio $r = 8$, and $T = 25$ °C. (Reproduced with permission from Ref. [20])

identifies the sol to gel transition; the crossover time approximates quite well the gelation time. Measuring G' and G'' at different frequencies as a function of reaction time gives a determination of the gelation time, t_g.

Figure 6.8 shows the changes in the storage and loss moduli, G' and G'', as a function of time for a silica sol (prepared from TMOS precursor) [20].

The response with the time of the moduli can be divided into three different regions. In the first one, there is only a viscous response; the gel behaves like a viscous material whose viscosity slowly increases with time because of the growth of the silica cluster dimensions; the gel does not have an elastic response. In the second region, which corresponds to the time close to the gel point, a sudden increase of the storage modulus, G', is observed; this is also accompanied by the appearance and fast rise of the loss modulus, G''. As we expect, close to the gel point, the material becomes mechanically resistant with an elastic response and very large viscosity. With the proceeding of the reactions beyond the gelation time (aging), the elastic modulus continues to increase, while the storage modulus, G', reaches a maximum.

A precise value of t_g can be obtained by the G'/G'' ratio. Figure 6.9 shows the case of a silica sol (TMOS). The G/G' ratio has been measured at different frequencies and plotted as a function of time, close to the gelation point. The intersection of the G''/G' curves at the different frequencies gives the gelation time; the measured value of t_g is independent on the frequencies.

The possibility of having a precise measure of t_g allows studying how the different preparation conditions affect the gelation time, in particular, the water/alkoxide molar ratio and the concentration of the precursor alkoxide.

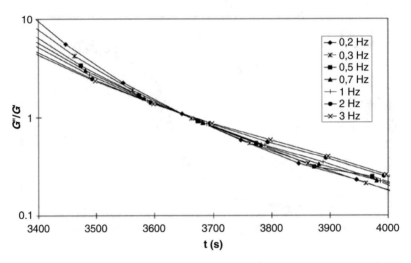

Fig. 6.9 Variation of the ratio of storage G' by loss G'' moduli near the sol-gel transition as a function of time at different frequencies for initial monomer concentration TMOS = 1.2 mol L^{-1}, hydrolysis molar ratio $r = 8$, and $T = 25$ °C. (Reproduced with permission from Ref. [20])

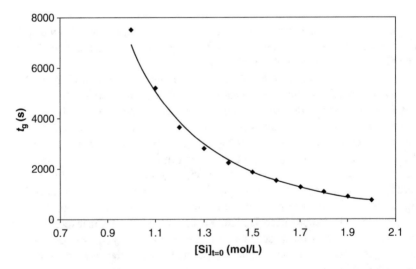

Fig. 6.10 Variation of the gelation time t_g as a function of the precursor TMOS concentration at constant water/alkoxide ratio, $r = 8$ and $T = 25$ °C. (Reproduced with permission from Ref. [20])

Figure 6.10 shows the change of gelation time, t_g, as a function of the alkoxide concentration, keeping constant the water/alkoxide ratio, r, and the reaction temperature, $T = 25°$. The gelation time decreases with the concentration of the alkoxide because a higher concentration favors a faster condensation due to the highest probability of the reactive species to encounter.

The same response should be expected if we keep constant the concentration of the alkoxide, but we change the water/alkoxide ratio. This is true, however, only in a limited range of r; we know in fact, from the discussion in Chap. 3 (see Fig. 3.5), that for values of r higher than 8, the dilution effect increases the gelation time. Fig. 6.11 shows the gelation time as a function of molar ratio, r, for three different concentrations of TMOS in the precursor sol. In the interval $(r = 2) - (r = 8)$, t_g decreases with the increase of r; the same trend is observed in all the sols regardless of the concentration.

It remains to correlate the t_g with the pH of the sol [21]. Figure 6.12 shows the change of storage modulus G' and loss modulus G'' as a function of time for a silica sol prepared using TEOS as precursor [22, 23]; the sols have been synthesized in acid, pH = 4, or basic conditions, pH = 9. The systems show some important differences: the kinetics of the sol-gel transition is much faster (up to almost two times) in the basic sol, but if we compare the values of G' and G'' at the same stage of the process, they are two or three times smaller in the basic sol. This means that even if the structure forms faster in the basic sol, it remains less interconnected with respect to the acid-catalyzed gel, which has a more rigid structure.

Some cautions have to be taken to evaluate these results if a full set of data for different pH values is not available because, in general, the gelation time also goes through a minimum, whose value depends on the synthesis and processing condition [24]. An example can be found in Fig. 6.13 for a water glass silica sol prepared at different pH values (pH = 2, 4, 6, 11); in the figure, the time dependence of G' and G'' from the aging time at different pH is shown. The intersection point of the G' and G'' curves gives the gelation time [23], whose values as a function of time are plotted in Fig. 6.14.

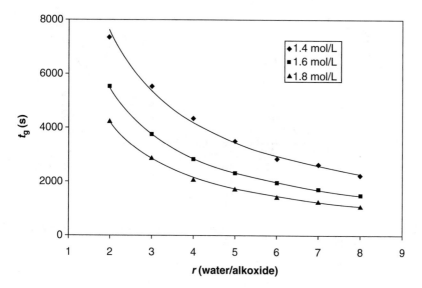

Fig. 6.11 Variation of the gelation time t_g as a function of the molar ratio r at different concentrations of TMOS and $T = 25$ °C. (Reproduced with permission from Ref. [20])

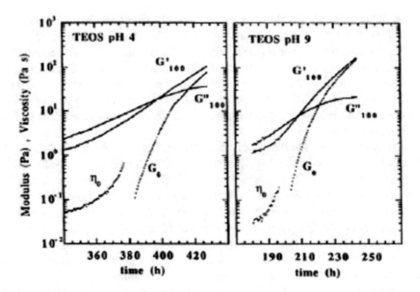

Fig. 6.12 Time variation of static viscosity η_0, elastic modulus G_0, storage modulus G' ($\omega = 100$ rad sec^{-1}), and loss modulus G'' ($\omega = 100$ rad sec^{-1}) or the acidic (pH = 4) and basic sol (pH = 9). TEOS/H$_2$O/EtOH = 1:10:6; gelation time at 16 °C for the acid sol, $t_g = 16$ days; and for the basic sol at 14 °C, $t_g = 8$ days. (Reproduced with permission from Ref. [23])

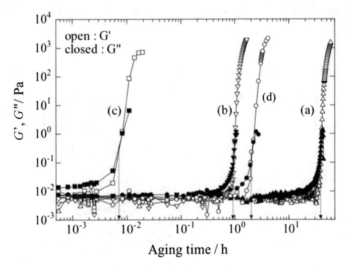

Fig. 6.13 Time dependence of G' and G'' of SiO$_2$ sols at pH = 2 (**a**), 4 (**b**), 6 (**c**), and 11 (**d**). (Reproduced with permission from Ref. [24])

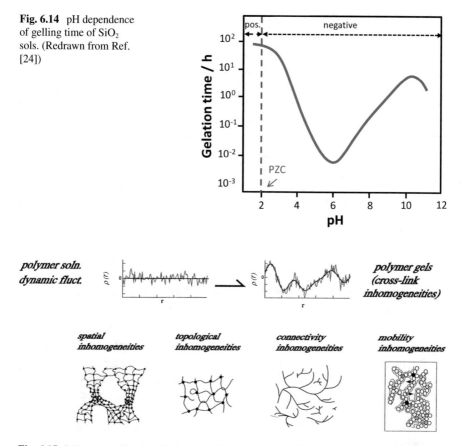

Fig. 6.14 pH dependence of gelling time of SiO$_2$ sols. (Redrawn from Ref. [24])

Fig. 6.15 Inhomogeneities in polymer gels. The upper figures illustrate the difference in concentration fluctuations between polymer solutions (left) and polymer gels (right). For polymer gels, in addition to thermal fluctuations, frozen inhomogeneities are superimposed. (Reproduced with permission from Ref. [28]

The gelation time is maximum when pH is lower than 2 (PZC positive) and at pH = 4 decreases to 1 h and drop up to 22 s at pH = 6; in basic conditions with pH = 11, the gelling time increases again to become 2 hours (Fig. 6.15). This trend is in agreement with the data reported by Iler for a silica sol of uniform particles [1] and with several other experimental results shown in the literature [25].

In general the pH-gelation time curves show a similar behavior even if they differ with the synthesis conditions. The higher temporary stability of the sol is observed at low pH and increases again at highly basic conditions (see also Fig. 3.4). The curve of gelation time shows three different regions, one at pH lower than 4, which is a metastable region where the sol has the highest gelation time and is relatively stable; an intermediate region, 4 < pH < 7, where condensation and aggregation is faster; and the third region at pH > 7 which is characterized by particle growth.

6.2.3   The Rheology of Sol-Gel Transition

In the previous paragraphs, we have briefly summarized some of the experiments which have been used to investigate the sol to gel transition in inorganic chemical gels by their rheological properties [26]. The condensation and growth process is associated with an enhancement of viscosity which increases with a power law until diverging near the gel point. The formation of a spanning interconnected cluster marks the gelation point and the appearance of a gel phase. The gel has an elastic modulus which rises close to the gelation point following a power law. The sol to gel transition characterizes, therefore, in rheological terms the change of the sol from a viscous liquid to an elastic material which is able to bear stress.

The process, even if the viscosity increases very quickly close to the gel point, happens without any physical discontinuity; this is the main reason why an accurate definition and measure of the gelation time is a rather complicated job.

6.3   Determination of Gel Point Through Dynamic Light Scattering

Time-resolved dynamic light scattering (*TRDLS*) is another method which has been applied to study the formation of a silica gel in situ [27, 28]. The technique has some advantages with respect to rheological methods because shorter time scales can be used to monitor the sol to gel transition and the measure does not interfere with the process.

On the other hand, the analysis of the experimental data requires some care because the clusters of different sizes that form during gelation create a signal scattering. A gelling system is not, in fact, a homogeneous material but rather a growing and "alive" system which evolves with the time. Several types of inhomogeneities can arise close to the gel point. Shibayama and Norisuye [27] have identified four different types of frozen inhomogeneities (Fig. 6.15).

In general, in a sol, only thermal concentration fluctuations appear whose average is zero. Gels, instead, contain also frozen concentration fluctuations which are due to the cross-linking of the polymer. The formation of a cross-linked network produces four types of inhomogeneities which are responsible of an anomalous scattering: *spatial inhomogeneities*; nonrandom spatial variations of cross-link density in a gel; *topological inhomogeneities* defects in the network, dangling chains, loops, and chain entrapment; *connectivity inhomogeneities* which depend on cluster size, distribution, and architecture of the polymer network; and *mobility inhomogeneities* variations of local degree of mobility by introduction of cross-links. These connectivity inhomogeneities can become significant just close to the sol to gel transition threshold. The frozen inhomogeneities give rise to strong scattering speckles (dots of lights scattered or reflected by the sample) which are used to identify the gel state and the sol to gel transition (Fig. 6.16).

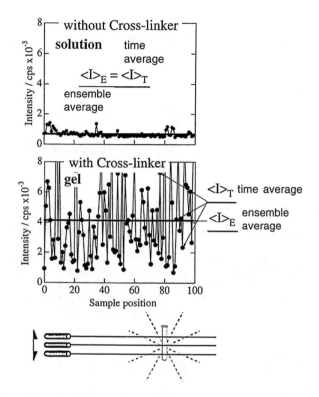

Fig. 6.16 Comparison of speckle patterns between polymer solution (upper) and polymer gels (lower). (Reproduced with permission from Ref. [28]

A gel state is characterized by the difference in the time-average scattering intensity $<I>_T$ and the ensemble-average scattering intensity $<I>_E$:

$$<I>_T \neq <I>_E \tag{6.8}$$

TRDLS has been applied to hybrid organic-inorganic [29, 30] and silica sols [31, 32] synthesized from TMOS to measure the gel point and obtain the scaling factor. TRDLS measures of acid and basic catalyzed TMOS sols have allowed identifying the gel point and the power law, as shown in Fig. 6.17. The two systems show a very different response with the reaction time; the average-scattered intensity of the acid sol does not change for the first 10 hours of reaction, to show a gradual increase after this time. Around 21 hours a strong fluctuation of the scattered intensity indicates the gelation of the system (dashed line in the figure) and a power law response. The basic system show, instead, a different variation of the scattered intensity with the reaction time (Fig. 6.17b). The intensity rises very quickly at the very beginning of the process and is followed by a gradual increase. This specific response is an indication that the two systems have very different kinetics of reaction, very slow at the beginning in the acid sol and fast, with the formation of "rigid" clusters that

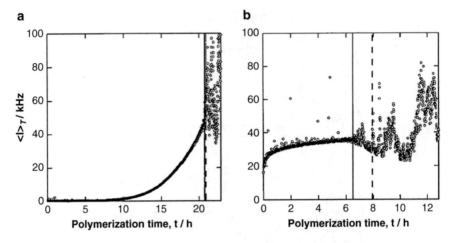

a

b

Fig. 6.17 Scattered intensity <I'>$_T$ variations (the time-average-scattered intensity observed at the fixed angle of 90°) of TMOS (concentration 5 wt %) during gelation process with (**a**) the acidic and (**b**) the basic catalysts. The solid and dashed lines indicate the time where strong fluctuations in <I'>$_T$ appeared and the time where a power law behavior appeared in the time-intensity correlation function (ICF), respectively. (Reprinted with permission from Ref. [32]

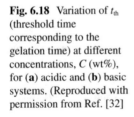

Fig. 6.18 Variation of t_{th} (threshold time corresponding to the gelation time) at different concentrations, C (wt%), for (**a**) acidic and (**b**) basic systems. (Reproduced with permission from Ref. [32]

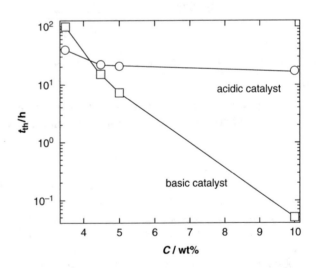

scatter the incident light, in the basic sol. After 6.5 hours of reaction, the strong fluctuations of the average-scattered intensity observed in the basic system, which are an indication of structural inhomogeneities, do not allow a determination of the gel point. This can be however identified through the power law response. At the gelation threshold, t_{th}, a power law behavior was observed for both acid and basic systems following a time-intensity correlation function, $g^{(2)}(\tau) - 1 \sim \tau^{n-1}$ [28].

The analysis by TRDLS has been also used to study the effect of the concentration of the precursor on gelation in the acid and basic sols (Fig. 6.18).

The TRDLS data show that the kinetics of gelation is much faster in basic than in acid sols. At concentrations higher than 5% in weight, the gelation time does not change so much in acid sols, while the basic systems are strongly dependent on the concentration. These results are another experimental demonstration of the different structures of the acid and basic sols. The power law exponent, n, which is the same critical exponent of viscoelasticity [29], differs in the two systems; it is $n = 0.73$ for the basic gels, and $n = 0.5$ for the acid gels. These values indicate that acid gels consist of long polysiloxane chains which are only weakly cross-linked, while a highly branched network forms the basic gels.

References

1. Iler CJ (1978) The chemistry of silica. Wiley, New York
2. Brinker CJ, Assink RA (1989) Spinnability of silica sols – structural and rheological criteria. J Non Cryst Solids 111:48–54
3. Ferry JD (1989) Viscoelastic properties of polymers, Third edn. Wiley, New York
4. Chang Q, Cerneaux S, Wang X, Zhang X, Wang J, Zhou J (2015) Evidence of ZrO_2 sol–gel transition by gelation time and viscosity. J Sol-Gel Sci Technol 73:208–214
5. Huggins ML (1942) The viscosity of dilute solutions of long-chain molecules. IV. Dependence on concentration. J Am Chem Soc 64:2716–2720
6. Einstein A (1906) A new determination of molecular dimensions. Ann Phys. 19: 289–306
7. Sakka S, Kamiya K (1982) The sol-gel transition in the hydrolysis of metal alkoxides in relation to the formation of glass fibers and films. J Non Cryst Solids 48:31–46
8. Sakka S, Kamiya K, Makita K, Yamamoto Y (1984) Formation of sheets and coating films from alkoxide solutions. J Non Cryst Solids 63:223
9. Brinker J, Scherer GWJ (1985) Sol → Gel → Glass: I. Gelation and gel structure. Non-Cryst Solids 70:301–322
10. Orgaz F, Rawson H (1986) Characterization of various stages of the sol-gel process. J Non Cryst Solids 82:57–68
11. Angulo-Olais R, Illescas JF, Aguikar-Pliego J, Vargas CA, Haro-Perez C (2018) Gel point determination of TEOS-BASED polymeric materials with application on conservation of cultural heritage buildings. Adv Condens Matter Phys 5784352:7
12. Sacks MD, Sheu RS (1987) Rheological properties of silica sol-gel materials. J Non Cryst Solids 92:383–396
13. Sacks MD, Sheu RS (1986) Rheological characterization during the sol-gel transition, in Science of ceramic chemical processing. Ed. by L. L. Hench, D. R. Ulrich, Wiley. New York
14. Jokinen M, Gyorvary E, Rosenholm JB (1998) Viscoelastic characterization of three different sol-gel derived silica gels. Colloids Surf A Physicochem Eng Asp 141:205–216
15. Cao XJ, Cummins HZ (2010) Structural and rheological evolution of silica nanoparticle gels. Soft Matter 6:5425–5433
16. Kikuchi S, Saeki T, Ishida M, Tabata K, Ohta K (2010) Sol-gel transition of acid silica sols produced by a Y-shaped reactor. Nihon Reoroji Gakkaishi 38:209–214
17. Winter HH, Chambon F (1987) Linear viscoelasticity at the gel point of a crosslinking PDMS with imbalanced stoichiometry. J Rheol 31:683–697
18. Winter HH, Mours M (1997) Rheology of polymers near liquid-solid transitions. Adv Polym Sci 134:165–234

19. Winter HH, Chambon F (1986) Analysis of linear viscoelasticity of a crosslinking polymer at the gel point. J Rheol 30:367–382
20. Ponton A, Warlus S, Griesmar P (2002) Rheological study of the sol–gel transition in silica alkoxides. J Colloid Interface Sci 249:209–216
21. Balkose D (1990) Effect of preparation pH on properties of silica gel. J Chem Technol Biotechnol 49:165–171
22. Devreux F, Boilot JP, Chaput F, Malier L, Axelos M (1993) Crossover from scalar to vectorial percolation in silica gelation. Phys Rev E 47:2689–2694
23. Kaide A, Saeki T (2014) Development of preparation method to control silica sol–gel synthesis with rheological and morphological measurements. Adv Powder Technol 25:773–779
24. Brinker CJ, Scherer GW (1990) Sol-gel science. Academic Press, San Diego
25. Knoblich B, Gerber T (2001) Aggregation in SiO$_2$ sols from sodium silicate solutions. J Non Cryst Solids 283:109–113
26. Winter HH (1987) Evolution of rheology during chemical gelation. Progr Colloid Polym Sci 75:104–110
27. Shibayama M, Norisuye T (2002) Gel formation analyses by dynamic light scattering. Bull Chem Soc Jpn 75:641–659
28. Richter S (2007) Recent gelation studies on irreversible and reversible systems with dynamic light scattering and rheology – a concise summary. Macromol Chem Phys 208:1495–1502
29. Aoki Y, Norisuye T, Tran-Cong-Miyata Q, Nomura S, Sugimoto T (2003) Dynamic light scattering studies on network formation of bridged polysilsesquioxanes catalyzed by polyoxometalates. Macromolecules 36:9935–9942
30. Norisuye T, Shibayama M, Tamaki R, Chujo Y (1999) Time-resolved dynamic light scattering studies on gelation process of organic-inorganic polymer hybrids. Macromolecules 32:1528–1533
31. Martin JE, Wilcoxon J, Odinek J (1991) Decay of density fluctuations in gels. Phys Rev A 43:858–871
32. Norisuye T, Inoue M, Shibayama M, Tamaki R, Chujo Y (2000) Time-resolved dynamic light scattering study on the dynamics of silica gels during gelation process. Macromolecules 33:900–905

Chapter 7
Probing the Sol to Gel Transition in the Gel Structure

Abstract The sol to gel transition in a bulk gel is a macroscopic event which can be followed using different analytical methods, and even if an accurate determination of the gelation time is always a difficult task, it is possible to obtain reliable measures especially by a combination of several techniques. The liquid phase of the gel remains, even after gelation, a peculiar chemical environment with a specific microviscosity. This evolves with drying, while diffusion inside the pores is always possible. The methods used to probe this microenvironment and sol-gel transition in thin films are the subject of this chapter.

Keywords Diffusion · Microviscosity · Drying · Time-resolved infrared spectroscopy

7.1 Diffusion During Gelation

Because of the relative freedom of movement of the molecules even after the gel point, molecular probes of different nature have been used to monitor the sol to gel transition [1]. The location within the system of the probe molecules is a critical point because the chemical environment is affecting the response of the probes. The peculiar microenvironments within the gel have been analyzed using molecular probes [2]. Different types of molecules can be used, which are optically active [3, 4] (spectroscopic probes) or electroactive [5, 6]; these probes can be embedded in the matrix by adding them into the sol or chemically bonded to the precursor. An example is the use of ferrocenic probes: one is the dimethylferrocenylmethyl(8-trimethoxysilyl)octylammonium bromide (FDA) (Fig. 7.1) which has been modified to bear a silicon alkoxide terminal group and therefore is able to attach to the silica species within the sol; the other is ferrocenylmethanol which, instead, is highly soluble in aqueous alcohol and is free to move within the gel.

These molecules allow studying the different electrochemical response of a silica gel (TMOS precursor) loaded with free or attached molecules. The kinetics of diffusion of the two probes gives direct information about the rigidity of the gel

© The Author(s), under exclusive license to Springer Nature Switzerland AG 2019
P. Innocenzi, *The Sol-to-Gel Transition*, SpringerBriefs in Materials,
https://doi.org/10.1007/978-3-030-20030-5_7

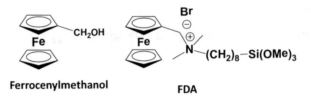

Fig. **7.1** The ferrocene probes, ferrocenylmethanol and dimethylferrocenylmethyl(8-trimethoxysily1)octylammonium bromide (FDA)

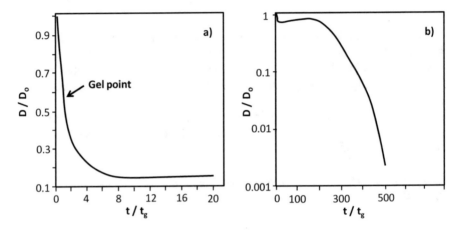

Fig. 7.2 Change of the reduced diffusion coefficient, D/D_0, in a base-catalyzed silica sol. (**a**) Ferrocenylmethanol, (**b**) FDA probes. (Redrawn from Ref. [7])

structure microenvironments and how this changes during gelation. Figure 7.2 shows how the diffusion coefficient changes (expressed in terms of the reduced diffusion coefficient, D/D_0, with D_0 the diffusion coefficient of the sol at t_g) as a function of t/t_{gel} for the two different probes in a base-catalyzed silica sol.

In the silica gel containing the ferrocenylmethanol "free" probe, the reduced diffusion coefficient does not change at the gelation point, even if the sol has solidified and has become rigid (Fig. 7.2a). The macroscopic structural change does not affect the mobility of the molecules which in the continuous liquid phase of the gel are still relatively free to diffuse. It has also been measured [8] that in a TMOS gel, the diffusion coefficient of NaCl is just a little smaller than in water, which gives a good idea of the fast transport that is still possible within a gel.

A different response is, instead, observed in the gel containing the bonded probes (Fig. 7.2b); in this case D/D_0 shows a sensitive decrease during gelation. The ferrocene probes are likely bonded to small silica clusters, and only when they attach to the spanning macromolecule, they are finally immobilized. This experiment shows very well the relative mobility of molecular species within the liquid phase of the gel and the aggregation process of the small oligomers in the sol. It is also clear that reaching the gel point is a macroscopic event which has not any special meaning at the molecular level, where silica clusters continue to diffuse and to grow also after gelation; this is the final stage of the sol-gel process which is indicated as aging.

Another family of probes which has been used to test the sol to gel transition is composed of fluorescent molecules. Fluorescent probes are very efficient tools in testing the changes in the surrounding chemical environment and have been used for measuring the variations of pH [9, 10], microviscosity, rigidity [11], and polarity [12] during the sol-gel process.

7.2   Microviscosity

The viscosity of the gel is a macroscopic rheological property whose quick increase close to the gel point indicates the transition from sol to gel. The molecules inside the gel, however, as we have just observed in the previous paragraph, are relatively free to move. This movement depends in large extent on the microviscosity inside the liquid phase which is important because it governs the optical response of guest molecules such as their mobility. Several works have been dedicated to understanding the microviscosity change during the sol to gel transition using local probes.

A method which has been largely applied to measure the local viscosity (microviscosity) is based on measuring the rotation of guest molecular probes. These techniques use fluorescent molecules whose optical response, due to the molecular rotation, is affected by the chemical environment. Fluorescence polarization, time-dependent fluorescence anisotropy [7], and fluorescence lifetime, to cite some of the most common, have been used for this purpose.

One of these experiments has been done by putting in correlation the fluorescence lifetime with the change in local viscosity during the sol to gel transition [13, 14]. If the fluorescence decay behavior is depending on the molecule rotations, a change in the chemical environment surrounding the probe will affect the fluorescence lifetime.

It is also possible to use different probes which have a specific affinity for the different environments within the gel. The molecular probes in Fig. 7.3 are an example: (**1**) is hydrophilic and is expected to stay in a water-rich environment, while (**2**), which is hydrophobic, would be located and preferentially stay closer to the silica network, because of water aggregates. The two probes can give complementary information about the changes in the local environment of the system during the sol to gel transition.

The change in viscosity as a function of reaction time obtained from lifetime data for the two probes is shown in Fig. 7.4; the samples are silica gel prepared from TEOS. The hydrophobic probe which is associated with the silica species senses a much higher viscosity than the hydrophilic probe which is in a microenvironment rich in water and has higher mobility. In the first minutes, the two probes have a similar trend with a quite fast increase in viscosity which follows an exponential growth. After the first stage, the viscosity does not change anymore around the

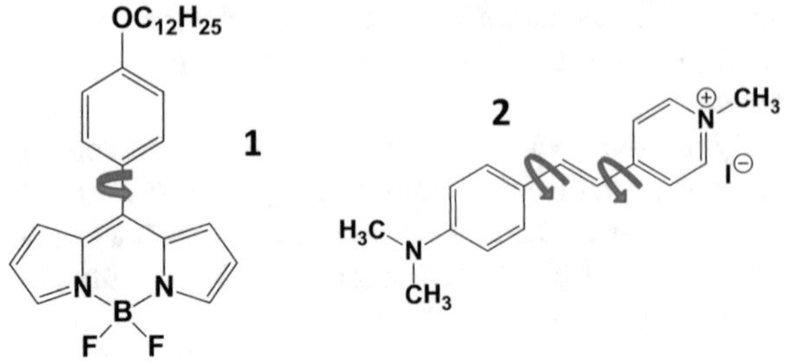

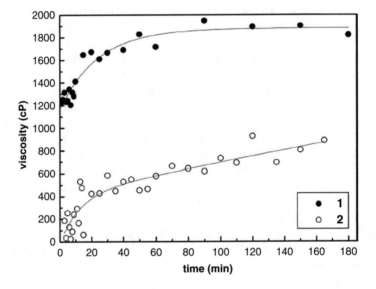

Fig. 7.3 Probe structures: (1) (4,4′-difluoro-4-bora-3a,4a-diaza-s-indacene) and (2) (4-(4-(dimethylamino)styryl)-N-methyl-pyridiniumiodine). The principal rotations giving rise to their fluorescence decay behavior are also shown in the figure. (Reprinted with permission from Ref. [15])

Fig. 7.4 Microviscosity for the sol to gel transition for TEOS calculated from lifetime data for the two probes. (Reprinted with permission from Ref. [15])

probe **2**, while a slow increase is still found for the hydrophilic probe. This increase in viscosity in the sol is due to the evaporation of water with time which has a consequence of the constant increase in viscosity.

Probing the microviscosity within the liquid phase of the gel can be done with high precision, and very nice results have been obtained to define local microenvironments within the pores; they have been identified by the different responses of fluorescent local probes [16].

An interesting group of photoprobes which has been used to monitor the effect of sol-gel transition within the local chemical environment is given by *rigidochromic* molecules which are complexes of the type ReCl(CO)$_3$L (L = bidentate diimine) [1]. In rigid media the emission wavelength of these molecules shifts to

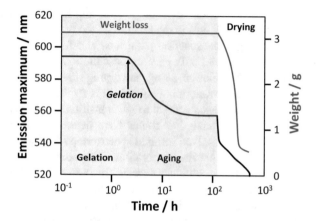

Fig. 7.5 Plot of the wavelength of the emission maxima of ReCl(CO)₃bpy in the aluminosilicate system as a function of processing time (lowest curve). The gelation point, the aging period, and the drying period are shown. The concomitant weight changes are shown in the upper curve. (Redrawn from Ref. [8])

higher energy (blue), and the quantum yield increases. In general, the local change in viscosity during gelation is high enough to produce a blue shift correlated to "rigidity" at the molecular level. These molecules have been used to monitor the sol to gel transition in an alumina system (Fig. 7.5). The shift of the emission wavelength maximum allows identifying three different stages in the process which corresponds to the three basic steps of the sol-gel process: gelation, aging, and drying. In the first step, which corresponds to the transformation of the alumina sol into a gel, no changes are observed.

The structural changes during this phase do not restrict the local mobility of the probe which does not sense any variation in the chemical environment. This in agreement with the results obtained using different probes that have shown no drastic changes at the gel point. In the second step which corresponds to the aging phase (the gel is left to react in a sealed vessel), the condensation reactions make the structure more and more "rigid" as the probe senses it. When the residual solvent is allowed evaporating (drying stage), a sudden blue shift of the emission wavelength is observed, which indicates the consolidation of the oxide structure and the loss of mobility of the probe. The simultaneous measurement of the weight loss shows that only in the last stage when water and alcohol evaporate, a drastic change is observed.

7.3 Probing the Sol to Gel Transition in Fast Evaporating Sols

Thin films are one of the most successful materials prepared via sol-gel processing; they are obtained by several deposition techniques, such as dip coating and spin coating, and with a variety of different compositions and structure such as from oxides, organic-inorganic hybrids, nanocomposites, and mesoporous layers [17]. The main difference with bulk gels is the kinetics of the process; the fast evaporation of the solvent changes the reaction pathways, and following the process in detail is not an easy task. While bulk gels are possible to separate, within certain limits of approximation, the gelation, aging, and drying stages, as we have just seen in the previous examples, in evaporating sol-gel films require only a few seconds. Identification of sequential stages and sol to gel transition remains therefore very difficult. Two main possible analytical routes have been used, fluorescent local probes [18] and in situ time-resolved infrared spectroscopy [19]. Another specificity of the film preparation is that the evaporation of the solvent creates a spatially resolved concentration of the chemical species; in the case of dip coating, for instance, interference fringes form as a function of the distance from the sol reservoir. The evaporation of the solvent in water-ethanol system is a dynamic process, and continuous exchange with the external environment is observed. In situ probing, using fluorescent molecules, has been used for silica and mesoporous films [15, 20]. Pyranine molecules, such as 8-hydroxy-1,3,6-trisulfonated pyrene, are sensitive to protonation-deprotonation effects and have been used to monitor the water/ethanol ratio change during dip coating of silica films (TEOS). Spatially resolved fluorescence spectra have shown that preferential evaporation of alcohol occurs and the solvent composition close to the drying line is enriched in water (80 vol% water).

An alternative technique for in situ monitoring of sol-gel reactions is FTIR time-resolved spectroscopy. This method has been applied to several evaporating systems, and the experiments can be performed using different setups such as attenuated total reflection (ATR) or an infrared microscope. Time-resolved FTIR analysis has been applied to ethanol [21] and water evaporating droplets [22, 23], mesoporous films [24–26], sol-gel films [27], hybrid materials [28], and tri-block copolymers [29, 30]. The technique has also been coupled with small-angle X-ray scattering (SAXS) to obtain a simultaneous time-resolved analysis of self-assembly and evaporation phenomena in mesoporous-ordered films [31]. The time scale of application is within seconds, and the technique can follow all the evaporative processes within this time range.

The preferential evaporation of alcohol during film deposition and the exchange of water with the external environment affect the kinetics of the sol-gel reactions and the gelation process. This is an effect that has always to be taken into account, and the control of conditions within the deposition chamber is, therefore, critical. Understanding the dynamic of the exchange of water with the external environment is, however, difficult because it is necessary discriminating between the contribution of water in the sol, including the amount generated as a by-product of condensation, and the absorbed water. To be able to study the process, the evaporation experiment has to be carefully designed to separate the different contributions. As a case study, a silica precursor sol has been synthesized using tetraethyl orthosilicate, deuterated water, and $SiCl_4$ as a catalyst, which avoids the addition of an acid source containing external protons. The absence of water added to the sol allows following the absorption of water from the external environment and the evaporation of the internal deuterated water; the two processes can be, therefore, analytically separated.

Monitoring with time the different FTIR absorption bands is possible to get simultaneous information about hydroxyl content (hydrolysis), -Si-O-Si- bond formation (condensation), deuterated water content (evaporation of water), water content (absorption of water), and alcohol content (evaporation of alcohol).

Figure 7.6a, c, e shows the 3D FTIR time-resolved absorption spectra (absorbance-wavenumber-time) of the three different series of evaporation experiments in the 3710–3015 cm^{-1} range. In this range wavenumber range with several overlapped bands is observed; they allow following the change of hydroxyls (3450 cm^{-1}) and absorbed water (3300 and 3490 cm^{-1}) during evaporation. The first column shows the change in absorbance intensity during evaporation in a false color scale. The experiment has been performed using a precursor sol containing different amounts of deuterated water, r = 1, 4, and 8. The sol with sub-stoichiometric water

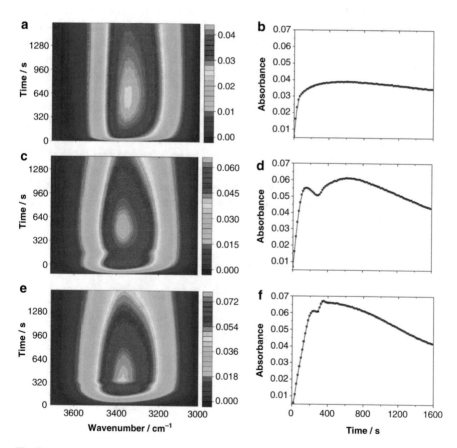

Fig. 7.6 3D ATR-FTIR time-resolved absorption spectra (absorbance-wavenumber-time), in the 3710–3015 cm^{-1} range, of the three of evaporation experiments corresponding to r = 1 (**a**), 4 (**c**), and 8 (**e**); the absorbance intensity is represented by a false color scale. The corresponding changes of absorption maxima as a function of time are also shown (**b**, **d**, and **f**, respectively). (Redrawn with permission from Ref. [24])

(r = 1) shows a progressive increase of OH species (Fig. 7.6a) up to a maximum at 540 s (Fig. 7.6b; after this time the OH content decreases. In the other two sols, which have stoichiometric and supra-stoichiometric amounts of D_2O, the OH content changes by following different and well-defined stages (Fig. 7.6c, e). These stages can be defined as it follows: a first stage characterized by a fast increase of hydroxyls content, a second shorter stage when the hydroxyls increase stops, a third phase when the content of OH groups enhances again and the last stage with a continuous hydroxyls decrease. These data can be coupled with the trend shown by the FTIR absorption bands of -Si-O-Si- and deuterated water to obtain a consistent picture of the phenomena observed during the sol-gel transition in fast evaporating systems [19]. Figure 7.7 shows how the FTIR bands evolve for the stoichiometric sample r = 4. The silanol and water bands are shown as a function of the change in intensity absorbance and the silica band as a function of wavenumber. Four different stages have been identified in the sol to gel transition during evaporation. The first one, which is the fastest one, is due to the hydrolysis of TEOS [32]. In this phase, the condensation is very small, while the hydroxyls quickly rise together with the absorption of water from the external environment. This is due to hydrophilic hydroxyls whose formation changes the nature of the sol, which was initially hydrophobic because of the unreacted TEOS.

The hydrolysis stage is followed by another short phase which is characterized by a decrease in hydroxyl content and a fast rise of condensation rate, while water absorption does not change very much. These data suggest that the condensation of silica requires a critical amount of silanols to be initiated and that the process is quite fast at this stage. The condensation of silanols produces a water release (in the experiment is deuterated water), which becomes available in excess. The depletion of silanols due to condensation also reduces the absorption of water from the external environment.

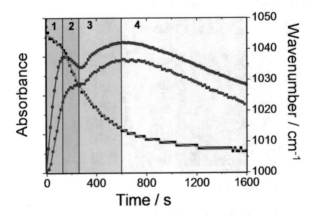

Fig. 7.7 Time-dependent curves (time-absorbance) of water (red) and OH (blue) superimposed with the time-dependent curve (time-wavenumber) of the -Si-O-Si- antisymmetric stretching (black). The three curves are referred to the stoichiometric sol r = 4. The lines are guides for the eyes. (Redrawn with permission from Ref. [24])

The following stage is characterized again by an increase of silanols and absorption of water but at a lower rate with respect to the first stage because since silica condensation has already progressed, water absorption and silanol formation reach a maximum at the end of this stage.

The final evaporation phase characterizes the fourth stage; absorbed water and hydroxyls decrease while the condensation rate is small and a gel finally forms. A similar response, with different kinetics, is observed in the sample containing suprastoichiometric amount of water, not, however, when water is present in substoichiometric conditions.

This means in, general, that complete hydrolysis is observed only in excess of water; if the water amount is low, the critical amount of silanols capable of triggering a fast polycondensation is difficult to reach. In this case, a smaller silica condensation is produced upon the sol evaporation.

The time-resolved data confirm that the hydrolysis and condensation reactions are not separated in time and take place simultaneously; they have, however, different reaction rates which mark the evaporation stages. The analysis shows that the critical evaporation phase is the second one; the sol to gel transition occurs in this short time interval and is the formation of a critical amount of hydroxyls which triggers the silica condensation and the formation of a quickly spanning network which finally forms a gel phase.

References

1. Dunn B, Zink JI (1997) Probes of Pore Environment and Molecule-Matrix Interactions in Sol-Gel Materials. Chem Mater 9:2280–2291
2. Dunn BS, Zink JI (2007) Molecules in Glass: Probes, Ordered Assemblies, and Functional Materials. Acc Chem Res 40:747–755
3. Keeling-Tucker T, Brennan JD (2001) Fluorescent Probes as Reporters on the Local Structure and Dynamics in Sol–Gel-Derived Nanocomposite Materials. Chem Mater 13:3331–3350
4. Dunn B, Zink JI (1991) Optical Properties of Sol-Gel Glasses doped with Organic Molecules. J Mater Chem 1:903–913
5. Audebert P, Griesmar P, Sanchez C (1991) Electrochemical Probing of the Sol-Gel-Xerogel Evolution. J Mater Chem 1:699–700
6. Audebert P, Griesmar P, Hapiot P, Sanchez C (1992) Sol-Gel-Xerogel Evolution investigated by Electroactive Probes in Silica and Transition-metal Oxide Based Gels. J Mater Chem 2:1293–1300
7. Narang U, Wang R, Prasad PN (1994) Bright FV, Effects of Aging on the Dynamics of Rhodamine 6G in Tetramethyl Orthosilicate-Derived Sol-Gels. J Phys Chem 98:17–22
8. Gits-Leon S, LeFaucheux F, Robert MC (1987) Mass transport by diffusion in a tetramethoxysilane gel. J Cryst Growth 84:155–162
9. Avnir D, Levy D, Reisfeld R (1984) The Nature of the Silica Cage as Reflected by Spectral Changes and Enhanced Photostability of Trapped Rhodamine 6G. J Phys Chem 88:5956–5959
10. McKieman JM, Yamanaka SA, Knobbe ET, Pouxviel JC, Parveneh S, Dunn B, Zink JI (1991) Luminescence and Laser Action of Coumarin Dyes Doped in Silicate and Aluminosilicate Glasses Prepared by the Sol-Gel Technique. J Inorg Organomet Polym 1:87–103
11. Xue M, Zink JI (2014) Probing the Microenvironment in the Confined Pores of Mesoporous Silica Nanoparticles. J Phys Chem Lett 5:839–842

12. Kaufman OVR, Avnir D (1986) Structural Changes along the Sol-Gel-Xerogel Transition in Silica as Probed by Pyrene Excited-State Emission. Langmuir 2:717–722
13. Hungerford G, Allison A, McLoskey D, Kuimova MK, Yahioglu G, Suhling K (2009) Monitoring Sol-to-Gel Transitions via Fluorescence Lifetime Determination Using Viscosity Sensitive Fluorescent Probes. J Phys Chem B 113:12067–12074
14. Hungerford G, Rei A, Ferreira MIC, Allison A, McLoskey D (2009) Application of fluorescence techniques to characterize the preparation of protein containing sol-gel derived hosts for use as catalytic media. Prog React Kin Mech 34:289–327
15. Huang MH, Dunn BS, Zink JI (2000) In Situ Luminescence Probing of the Chemical and Structural Changes During Formation of Dip-Coated Lamellar Phase Sodium Dodecyl Sulfate Sol–Gel Thin Films. J Am Chem Soc 122:3739–3745
16. McKiernan J, Pouxvie J-C, Dunn B, Zink JI (1989) Rigidochromism as a Probe of Gelation and Densification of Silicon and Mixed Aluminum-Silicon Alkoxides. J Phys Chem 93:2129–2133
17. Faustini M, Boissière C, Nicole L, Grosso D (2014) From Chemical Solutions to Inorganic Nanostructured Materials: A Journey into Evaporation-Driven Processes. Chem Mater 26:709–723
18. Nishida F, McKiernan JM, Dunn B, Zink JI (1995) In Situ Fluorescence Probing of the Chemical Changes during Sol–Gel Thin Film Formation. J Am Ceram Soc 78:1640–1648
19. Innocenzi P, Malfatti L, Carboni D, Takahashi M (2015) Sol-to-Gel Transition in Fast Evaporating Systems Observed by in Situ Time-Resolved Infrared Spectroscopy. ChemPhysChem 16:1933–1939
20. Franville AC, Dunn B, Zink JI (2001) Molecular Motion and Environmental Rigidity in the Framework and Ionic Interface Regions of Mesostructured Silica Thin Films. J Phys Chem B 105:10335–10339
21. Innocenzi P, Malfatti L, Costacurta S, Kidchob T, Piccinini M, Marcelli A (2008) Evaporation of ethanol and ethanol-water mixtures studied by time-resolved infrared spectroscopy. J Phys Chem A 112:6512–6516
22. Innocenzi P, Malfatti L, Piccinini M, Marcelli A, Grosso D (2009) Water evaporation studied by in situ time-resolved infrared spectroscopy. J Phys Chem Chem A 113:2745–2749
23. Innocenzi P, Malfatti L, Piccinini M, Sali D, Schade U, Marcelli A (2009) Application of Terahertz Spectroscopy to Time-Dependent Chemical-Physical Phenomena. J Phys Chem A 113:9418–9423
24. Innocenzi P, Kidchob T, Bertolo JM, Piccinini M, Guidi MC, Marcelli A (2006) Infrared Spectroscopy as an In-Situ Tool to Study the Kinetics of Processes Involved in Self-assembly of Mesostructured Films. J Phys Chem B 110:10837–10841
25. Falcaro P, Costacurta S, Mattei G, Amenitsch H, Marcelli A, Guidi MC, Piccinini M, Nucara A, Malfatti L, Kidchob T, Innocenzi P (2005) Highly ordered 'defect-free' self-assembled hybrid films with a tetragonal mesostructure. J Am Chem Soc 127:3838–3846
26. De Paz-Simon H, Chemtob A, Croutxé-Barghorn C, Rigolet S, Michelin L, Vidal L, Lebeau B (2013) Block Copolymer Self-Assembly in Mesostructured Silica Films Revealed by Real-Time FTIR and Solid-State NMR. Langmuir 29:1963–1969
27. De Paz-Simon H, Chemtob A, Croutxé-Barghorn C, Le Nouen D, Rigolet S (2012) Insights into photoinduced sol–gel polymerization: an in situ infrared spectroscopy study. J Phys Chem B 116:5260–5268
28. Innocenzi P, Figus C, Takahashi M, Piccinini M, Malfatti L (2011) Structural Evolution during Evaporation of a 3-Glycidoxypropyltrimethoxysilane Film Studied in Situ by Time Resolved Infrared Spectroscopy. J Phys Chem A 115:10438–10444
29. Innocenzi P, Malfatti L, Piccinini M, Marcelli A (2010) Evaporation-Induced Crystallization of Pluronic F127 Studied in Situ by Time-Resolved Infrared Spectroscopy. J Phys Chem A 114:304–308
30. De Paz-Simon H, Chemtob A, Croutxé-Barghorn C, Rigolet S, Michelin L, Vidal L, Lebeau B (2013) Block copolymer self-assembly in mesostructured silica films revealed by real-time FTIR and solid-state NMR. Langmuir 29:1963–1969

31. Innocenzi P, Malfatti L, Kidchob T, Costacurta S, Falcaro P, Piccinini M, Marcelli A, Morini P, Sali D, Amenitsch H (2007) Time-Resolved Simultaneous Detection of Structural and Chemical Changes during Self-Assembly of Mesostructured Films. J Phys Chem C 111:5345–5350
32. Rubio F, Rubio J, Oteo JL (1998) A FTIR study of the hydrolysis of tetraethylorthosilicate (TEOS). Spectrosc Lett 31:199–219

Conclusion

At the end of this brief overview of the sol to gel transition in inorganic systems, a final summary can be outlined. The first studies on sol-gel processing have been very much focused on the possibility of obtaining bulk gels and oxides and have described the method as an alternative route to obtain glass from low temperature. Within the time this first idea has been practically abandoned, sol-gel inorganic chemistry has become something different; nowadays it is an almost ubiquitous process in nano-chemistry which is used to prepare a variety of different materials in the form of films, membranes, nanoparticles, aerogels, mesoporous and microporous materials, self-assembled materials, etc.

This change of perspective has brought to an unexpected success of inorganic sol-gel chemistry as a very popular tool for nanoscience; at the same time, part of the interest in the basic chemistry of the process has been lost which has also made many people unaware of the fundamental scientific background. We have seen that the complexity of the inorganic sol-gel chemistry represents a natural limitation to our capability of giving very general descriptions of the process. This is clearly reflected by the failure (classic model) or partial failure (percolation models) of theories to give an adequate description and prediction of the transition which reflect the main experimental findings.

The structure of silica units allows considering the condensation reactions as a polymerization process, and this is also true if functional organic substituents are present in the precursors via not-hydrolyzable Si-C bonds. In the case of transition, metal alkoxides and mixed oxides are still the chemistry which governs the process. On the other hand, the sol to gel transition is a critical phenomenon and is observed in organic polymers as such as more in general in colloidal systems. In inorganic systems, the triggering event which eventually drives the system to gelation is hydrolysis; after this first step, a continuous sequence of chemical reactions and aggregation gives rise to the formation of a spanning cluster and finally gelation. This event is not characterized by any specific macroscopic change and this makes a determination of the gelation time quite difficult. The gelation of the system is,

© The Author(s), under exclusive license to Springer Nature Switzerland AG 2019
P. Innocenzi, *The Sol-to-Gel Transition*, SpringerBriefs in Materials,
https://doi.org/10.1007/978-3-030-20030-5

however, accompanied by the divergence of macroscopic properties, such as viscosity; this is what is generally used to get an experimental evaluation of the sol to gel transition. Viscoelasticity experiments have been used as one of the main tools to identify the gel point; the change of mechanical properties, which means the rise of an elastic response under shear stress, characterizes the formation of a gel. In fast evaporating systems, such as sol droplets or thin films, the things are much more complicated, and identification of the sol to gel transition is possible only via in situ time-resolved FTIR.

The set of experiments which have been dedicated to understanding the sol to gel transition in inorganic and hybrid systems have also underlined that there is a strong dependence on the synthesis conditions and that the nature of oxide oligomers governs much of the aggregation phenomena and the gel transition.

Another interesting point to underline is that the sol-gel transition, which is a macroscopic event, is not reflected in any direct change within the residual sol, and several properties, such as diffusion in the liquid phase, are not affected by gelation.

The complexity of the phenomenon in the case of hybrid and inorganic systems does not allow us giving a complete and exhaustive description; still so much has to be understood, at least in terms of our capability of matching between experimental results and theory. The complexity is damnation but a wonderful challenge.

Index

A
Aerogels, 5, 97
Alkoxides, 8–18, 21, 25–30, 32, 40, 50, 56, 59, 62, 75–77, 85, 97

B
Bethe tree, 40, 43
Bond percolation, 44, 47, 48, 50
Bridged polysilsesquioxanes, 15, 16
Bulk gel, 5, 68, 90, 97

C
Cayley tree, 40
Classic models, viii, 42, 49, 55–60, 97
Clusters, 21, 29, 31, 33, 41, 42, 44, 45, 47–49, 51, 53, 67–69, 75, 80, 81, 86, 97
Colloids, vii, 2, 4
Condensation, 7, 10, 14, 16, 18, 21, 23, 25–33, 36, 37, 40, 42, 50, 53, 56, 58–61, 64, 72, 76, 79, 80, 89, 90, 92, 93, 97
Cyclization, 32, 33, 35, 56, 59–63

D
Diffusion, 35, 85–87, 98

E
Einstein relations, 69
Electrophilicity, 17
Esterification, 22
Evaporation, 88, 90, 91, 93

F
Ferrocene, 86
Films, 5, 68, 89, 90, 97
Flory, 41
FTIR, 90–92, 98

G
Gelation, viii, 3–5, 15, 31, 33, 35, 41, 42, 44, 50–53, 55–65, 67–69, 72–80, 82, 83, 85–87, 89, 90, 97, 98
Gelation time, vii, 24, 28, 33, 34, 51, 53, 59, 60, 62, 68, 72–80, 82, 83, 97
Gel point, 4, 6, 41, 42, 47, 50, 51, 53, 67–83, 85–87, 89, 98
Gels, vii, 1, 7, 21, 39, 56, 67, 85, 97

H
Huggins equation, 69
Hydrolysis, 7, 8, 10, 15, 18, 21–32, 36, 40, 53, 56, 59–61, 64, 72, 75, 76, 91, 92, 97

L
Light scattering, 80–83
Loss modulus, 74, 75, 77, 78

M
Methyltrimethoxysilane (MTES), 12
Microenvironments, 68, 85–88
Microviscosity, 87–89
Miscibility, 22

© The Author(s), under exclusive license to Springer Nature Switzerland AG 2019
P. Innocenzi, *The Sol-to-Gel Transition*, SpringerBriefs in Materials,
https://doi.org/10.1007/978-3-030-20030-5

N
NMR, 52, 59, 62
Non-random cyclization, 59, 60

O
Oligomerization, 62–65
Oligomers, 37, 55–65, 86, 98
Organic-inorganic hybrids, 4, 44, 90

P
Percolation, vii, viii, 43–53, 59–61, 68, 72, 97
Percolation threshold, 45, 46, 48–50, 53
Point of zero charge (PZC), 23–24, 26, 79
Polycondensation, 3, 11, 13, 26, 28, 29, 57, 58, 60, 70, 93
Polymerization, 3, 4, 11, 14, 40, 41, 44, 57–59, 97

R
Random branching, vii, 55, 62
Reduced viscosity, 68
Relative viscosity, 69
Rheology, 80

S
Scaling, 48–49, 51, 52, 81
Silica, viii, 4, 8, 23, 47, 56, 69, 85, 97
Silsesquioxanes, 14, 15, 33
Site percolation, 44–47
Sol, vii, 1, 7, 21, 41, 56, 67, 85, 97
Sol-gel transition, vii, viii, 4, 5, 56, 76, 77, 80, 92, 98

Spectroscopic probes, 85
Spectroscopy, 90
Storage modulus, 73–75, 77, 78
Structures, vii, 2, 4, 8, 11, 13–15, 26, 29, 31, 32, 35, 36, 39, 41–43, 47, 56, 57, 59, 62–64, 67–70, 72, 73, 77, 83, 85–93, 97

T
Tetraethyl orthosilicate (TEOS), 8–11, 21–24, 42, 59–63, 70, 73, 77, 87, 88, 90, 92
Tetramethylorthosilicate (TMOS), 8–10, 34, 51–53, 56–58, 62–65, 75–77, 81, 82, 85, 86
Time-resolved infrared spectroscopy, 90

U
Universality, 48–49

V
Viscoelasticity, 68, 83, 98
Viscosity, 4, 8, 10, 51–53, 67–73, 75, 78, 80, 87, 89, 98

W
Water glass, 17, 18, 77

X
Xerogel, 5, 31

Printed in the United States
By Bookmasters